SHAPING DESTINY

AMBITION – DETERMINATION – LEADERSHIP

KANWAL SETHI

◆ FriesenPress

Suite 300 - 990 Fort St
Victoria, BC, V8V 3K2
Canada

www.friesenpress.com

Copyright © 2016 by Kanwal Sethi
First Edition — 2016

All rights reserved.

No part of this publication may be reproduced in any form, or by any means, electronic or mechanical, including photocopying, recording, or any information browsing, storage, or retrieval system, without permission in writing from FriesenPress.

ISBN
978-1-4602-9376-8 (Hardcover)
978-1-4602-9377-5 (Paperback)
978-1-4602-9378-2 (eBook)

1. BIOGRAPHY & AUTOBIOGRAPHY, PERSONAL MEMOIRS

Distributed to the trade by The Ingram Book Company

To my loving wife,
Aruna for her steadfast support throughout

Table of Contents

Foreword vii

Introduction 1

Early Kenya and the British Empire 8

Decolonization 16

Princess Elizabeth becomes Queen 21

Migrating from India to Kenya 23

Move to Nairobi 35

A career in the army 45

The King's African Rifles: enlisting 52

Mons Officer Cadet School, Aldershot, UK 55

Regular Commissions Board, Westbury, UK 58

The Royal Military Academy, Sandhurst, UK 61

Fireside chat: Jallianwala Bagh 67

Horseback riding 72

Friendships at Sandhurst 75

Hitchhike to El Alamein, Egypt 81

Hitchhiking, Europe 90

Westminster Abbey, Colour Party, London 93

France—Exercise Normandy scholar and Moulin Rouge ...96

Iceland, in a fishing trawler ...100

Sovereign's Parade, Sandhurst ...107

Return to Kenya ...110

The Shifta insurgency ...112

Deployment in the NFD ...114

Operation Merti ...125

Knowing your men ...129

Military Training School, Lanet, Kenya ...143

Training officer cadets ...145

Defence headquarters, strategies and planning ...150

Rising from the dead ...152

Aid to civil power, Nyanza floods, 1967 ...155

Briefing Kenyatta ...160

Destiny by accident—auto dealership ...165

New Kenya Battalion, 7th Kenya Rifles ...171

Assisting in anti-poaching ...175

Agriculture Society shows ...179

Sports ...181

East African Coronation Safari ...183

Deployment in Garissa ...185

Garissa, watermelon farm ...188

Staff College, Camberley, UK 192

Destiny, unplanned; Libya 195

Kenya, coup? 198

Defence headquarters—staff duties 206

Write a book? 210

Post-army 215

Ghosts—believe or not 222

Burglary—house broken in 224

General Manager, of a local enterprise 226

Travelling surprises, Yemen 231

Kenya, future 235

Canada—a land of milk and honey 237

Starting a business, self-employed 239

Travels with Jacques Chouinard 256

Voluntary work, sports in Saint Bruno, Quebec 259

Voluntary work—Canadian Forces College 261

Canadian Forces College, Toronto, the Sethi Inukshuk Award 263

Canada, a home forever 266

Conclusion 268

References and Bibliography 273

FOREWORD

By Dr. Peter Foot
Emeritus Professor of Defence Studies,
Canadian Forces College, Toronto;
Associate Fellow (formerly Academic Dean), Geneva Centre for
Security Policy, Geneva.

Kanwal Sethi has two good reasons to share his story. The first is that he has lived an uncommon life in extraordinary times. For most of us, for most of our lives, the world has been artificially dominated by the transient contingencies thrown up by the superpower rivalry between the United States of America and the Union of Soviet Socialist Republics. We call that period the Cold War. For those involved in the news media, senior politics, the security services, and international discussions, this was the dominant narrative of the times. Two capitals, Washington and Moscow, with their not always consistent or wholly understandable priorities, dominated the agenda.

Without in any way seeking to do so, Kanwal Sethi, with this account of his life within his times, has shown that superpower agenda to have been less significant than the one that developed in parallel, and which led to many (and many more) profound consequences, the conclusions of which are still far from clear. The decolonization process dominates the twenty-first century to a greater extent than the late twentieth century superpower rivalry. That largely liberating process had many origins, not least the naval battle of Tsushima in 1905, which saw the first defeat of a European empire by an Asiatic power.

Russia failed to understand the progress Japan had made in education, technology, modernity of military power, and in its sense of control and management. Great Britain's narrow escape in the Boer War was another harbinger of change. Following the defeat of Japan in 1945, it was simply not possible for the former great powers of Europe to reassert their authority. Their time had passed.

France's agonies in Algeria and Indochina, as examples among many, are eloquent testimonies to that. The future would be shaped by those who were previously imperial subjects, described—and it's impossible not to be embarrassed by this today—by Rudyard Kipling as "lesser breeds without the law."

Kanwal Sethi's origins give the lie to all of that. His was part of an Indian family in an African setting: at some point, some nameless British official facilitated the movement of his family from its Indian subcontinent village to an East African setting where the Empire had, apparently, greater need for its labour. Kanwal's father responded magnificently to this challenge; Kanwal's story is the consequence of that. It happened to be paralleled by constituent parts of the European empires demanding to be given the dignity of independent statehood following the end of European global dominance that the outcome of World War II had made clear to all but the possessors of imperial power. Apart from anything else, what was the teaching of the liberal empires, *la mission de la civilisatrice*, supposed to produce?

There are complexities in all of this, of course. The post-independence "Africanization" program in Kenya, as in other places, could not be helpful to the Sethis—even though the consequences were personally less dramatic than for Ugandan Asians under Idi Amin.

Serving as an Indian in an African army at a senior level during a foundational period of nation-building was not—and could not be—straightforward. Seeing the British Empire and its Commonwealth legacy, for all of their faults, as primarily constructive, does not make for many friends in an independence-minded, newly created African state. The welcome that Canada offered Kanwal Sethi was both natural and entirely true to Canada's internationalist traditions within the Commonwealth.

Looking back through Kanwal Sethi's account, one discerns the conditions that fate determined for him. Kanwal, the child of India born into an aspirational family that happened to find itself in Kenya at a moment of historical turbulence and fading imperial power, possessed just the gifts necessary to make the most of that fate. Hamlet says, "There's a divinity that shapes our ends, Rough-hew them how we will." It is a semi-fatalistic idea: destiny has the major say in our outcomes, but we do have the ability to make the most—or otherwise—of what opportunities we are given. Reading Kanwal Sethi's memoir, one cannot but conclude that his parents—unintentionally almost bit parts in this drama—would be immensely proud. And Canada can be proud that its culture—that of the Commonwealth and its multifarious progeny—gave their son such a distinguished opportunity, so splendidly realized. That is the second reason for sharing this story.

INTRODUCTION

What does it take to shape a destiny? What influences a choice and selection of a career path? Is it a social environment, opportunity or lack thereof, financial or intellectual resources, ambition, initiative, creativity, determination and relentless drive to succeed against odds, obstacles and disappointments along the way, or role models to emulate? Perhaps it is a combination of all these factors and the ability to foresee, identify a goal, recognize limitations, build upon all kinds of resources with the confidence that any reasonable goal is achievable by remaining focussed.

I have been fortunate in having witnessed many events of a historical nature and, in the course, having met people who were larger than life and whom I admired for their extraordinary personalities. I drew inspiration from them and their influence on me has been everlasting. For some people, opportunities just appear miraculously or are handed down, on a silver platter as it were, with little or no glimpse into what it takes to earn a day's living, nurture ambition, set a life goal, or manage a regular day's affairs amidst limited resources. For the vast majority of us in an ever-more-competitive world, opportunities are searched for with much effort and some appear achievable while others are only hoped for and remain a distant mirage.

Every story of life's journey is different. Each one of us at some stage in our life, often in our teen years, establishes an aim to aspire to and a goal to fulfil. In *Shaping Destiny*, I have tried to describe my adventurous journey as a native-born Kenyan of Indian ancestry, from growing up in colonial times to making a home in Canada. Over this time, my varied and vast experiences exposed me to a racially segregated environment and a historically changing tide with limited opportunities. This compelled me to eventually put roots down in a new, strange land of plenty with unlimited opportunities to reach for: Canada.

Whenever I have shared my enormous experiences at home over the dinner table, with colleagues at work, or in a social environment, I have been intrigued by how much enthusiasm I have been able to generate.

The audience has been attentively listening to the unique adventures that I have been fortunate to have experienced. Many across a wide spectrum suggested that I should write my story as an example of what can be achieved with perseverance.

Lack of financial resources was always a pressing concern for me, since I came from a one-income, civil-servant family that had to make do with sparse affordable living and curtailed extravagant excesses. The blatant, immoral, affluent, and, in many respects, quite wasteful consumerism of today was not in our vocabulary. The seven of us lived within the income of a single breadwinner, and still had to make provisions for saving for a rainy day.

The present-day safety nets of free medical care, free schooling, unemployment insurance, and ease of job-hopping did not exist then. Further, the extended family—including close friends of the family with whom bloodlines were blurred—expected to be assisted materially and financially without having to ask and lower their self esteem. Such a lifestyle, ingrained from an early age, bred compassion, regard for others, and a willingness to share what little one could. It also laid a strong foundation of values and conditioned our character.

Along the way, numerous obstacles had to be overcome and prejudices overlooked. The belief was that networking and relying on others to open doors was the key to get ahead. Remaining cocooned in a box, isolated and complaining, would mean loneliness and negativity with the world whizzing by. Hard work, integrity, loyalty, and honesty are other key factors that identify a person and reflect on his or her character. These factors are judged by others who, in turn, form opinions—and it is amazing how easily help comes, often from unexpected sources.

I have never been deterred by rank and authority, but have always displayed understanding and respect for status in the belief that every one of us deserves a spot in the social and workplace hierarchy, no matter our circumstances.

We lived in government housing, which was an employment perk for father right up to my teen years. A desire to own a house one day was always there as life's ultimate goal, but it was not affordable on a meagre civil servant's salary, with other priorities such as fulfilling the basic needs of a family trumping every other priority. And there were lots of priorities and juggling them in whatever manner was pointless: the end result was always the same. Father managed to build and own our first house in Nairobi when I was in my teen years, but I would often overhear conversations at the dinner table about how tight we were on meeting monthly mortgage payments.

We were forced to rent out a part of the house to generate income to meet our ever-increasing living costs. But, most importantly, even against such odds, the atmosphere at home was full of joy, happiness, and constant prodding to excel at school, as a good sound education was acknowledged as the key to the success that would one day lead us to rise in upward social mobility. We never lacked the basic necessities or felt that we were missing out on a luxury of some nature or needed to keep up with anybody. We were comfortable as a family unit, and participated in community gatherings with pride. We enjoyed immense respect and were looked upon as pillars of strength, thereby providing hope to others.

I began my life with an abundance of love and inspiration, and with the confidence that anything was achievable. But I also understood that achievement required preparation. This meant acquiring a good education and undertaking a relentless pursuit of an ambition. By sheer luck and probably historical circumstances, several personalities became role models to whom I looked for inspiration. A few of them even volunteered guidance of their own volition with the expectation that I would consider their advice part of a natural cycle of evaluating options. Achieving success in established surroundings with familiar people is fine and relatively effortless, but crossing frontiers voluntarily into distant, unknown lands is totally different. Such a transition in search of a new home became a necessity with the realization that, with resourcefulness, support, and adjustment, better days lay ahead.

This realization is comparable to the experience of our folks embarking on a similar journey in the hope of a better tomorrow from the Indian sub-continent to Africa against every imaginable obstacle and in the company of great risks, uncertainty, and adverse circumstances in the early 1900s.

I am in debt to many individuals for their guidance and support, and the list of those who showed me the way and shaped my future is long. I am sure to miss out on many who deserve my recognition. However, a special mention is due to General Miles Fitzalan Howard, my first brigade commander of 70th Brigade, the King's African Rifles, who later inherited the dukedom of Norfolk; Sir Anthony Duff, who was the British high commissioner in Kenya in the early 1970s and later was appointed to lead the ultra-secretive MI5 in the United Kingdom; Murray Cook, the Canadian high commissioner in Kenya in the early 1960s who later headed human resources at External Affairs in Ottawa; Major General Joe Ndolo, the first native chief of defence staff in Kenya in the late 1960s under whom I served as military assistant; and General Jacques Chouinard of the Canadian army, whom I befriended in Canada and with whom I travelled extensively in Africa to promote business after moving to Canada.

Way back in the 1970s, Sir Anthony was the first to suggest that, one day, I should write my story, as it was unique and transcended historical milestones. And he was later to question—in the early 1990s by which point he had retired—what I had done about my story. Such courteous reminders by someone in the highest corridors of global power have been most humbling.

My biggest debt of gratitude is to my loving wife of forty-seven years, Aruna, who has stood by me as a rock, sharing and supporting every decision I've made. On several occasions when the going was difficult and uncertainty abounded, she never wavered, but gave hope and always looked at the worst of a situation positively. As with any military wife of that era, she participated in the welfare of my soldiers by visiting their families when I was away on deployment in the NFD. She attended passionately to whatever issues were brought to her notice and followed through as best as she could.

Officers' wives played key roles in soldiers' welfare, and were seen as an extension of the officers' authority. Beyond holding a full-time day job, Aruna happily managed a young, growing family with love, warmth, and closeness. She fully participated in the school activities of my two children, Meera and Sanjay. She thus relieved me of this important responsibility so that I could concentrate on building my career whilst in the army and, later, when I ventured into private business. Her boundless energy has been extraordinary as evidenced by how my two children grew, acquired university educations of their choosing, got married, and are now raising families of their own.

Aruna enriched my family with extraordinary talents. I stopped buying lottery tickets after meeting her, as she has been my jackpot win. Her devotion and selfless support to my immediate and extended family have been remarkable. Soon after our marriage in 1970, she took over the leadership of the family clan and provided direction. With her natural caring instincts, jovial personality, and infectious smile, she easily overrode the daughter-in-law designation and was simply a daughter.

Throughout, she ensured that every family member bonded closely with each other. She took care of my aging parents as well as her own, and did so beyond the call of a nurse with an abundance of love and respect, right up to their passings. She has been a role model as a mother-in-law and grandmother. She has always ensured that birthdays, anniversaries, and other celebrations within the family are observed with pride and happiness, thereby fostering cohesiveness and providing strength and a deep sense of attachment to all.

Aruna relentlessly pushed and encouraged me to start writing my story, and guided me regarding the key elements on which I might focus. A special thank you is due to Dr. Peter Foot, professor emeritus at the Canadian Forces College in Toronto and formerly head of the Geneva Centre for Security Development, for reviewing my several drafts, for suggesting the book format and title, *Shaping Destiny* (an excellent pick for the way it perfectly captures the theme), and for writing a foreword amidst his busy schedule.

Shaping Destiny reflects in us all in what it takes to live and grow in an era of changing times, political upheavals, segregated social environments, and factors that influence one's selection of a career path. This is a journey of risks and extraordinary adventures in my search to better myself. My goal throughout has been to provide hope and opportunities for my family, the next generation, and many others, and to thereby shape a destiny.

For the subject matter in *Shaping Destiny*, I have expanded on the many notes I made in a diary that I fortunately held onto. For those who have passed on, I relied on my memory and relived events as best as I could recall. For historical references that indicate dates and details of specific events, I consulted Wikipedia, the internet, and other publications listed at the end.

I have tried to portray my story in the context of evolving historical surroundings as best and accurately as I could. I am responsible for the content and, if readers find any inaccuracies, I ask for their understanding as my attempt is not to mislead or make false representations.

My journey over the years has been an enormous success. I started with an empty pocket, opting to walk a distance to save on a meagre bus fare, and eventually leading to the establishment of a thriving successful business. Hopefully, this will give a head start and enable members of the next generation to build their ambitions. *Shaping Destiny* is my account of the wonderful, exciting path I have followed from humble beginnings.

EARLY KENYA AND THE BRITISH EMPIRE

As a backdrop to the Kenya of today, the country has gone through several transformations via migration, conquest, religion, and shaky alliances among warring tribal groups. The lifestyle of Kenyans has varied from pastoral to nomadic, with tribal territories joining in over time and forming countries with borders that were, in many cases, identified by colonial powers as independent countries.

Africa

The first people to migrate and settle in Kenya were indigenous African communities—the Cushitic, Nilotic, and Bantu—some 2,000 years ago. The Cushitic arrived from the north via Sudan and Ethiopia, bringing hunting skills with them. Their descendants are the El Molo, Somali, and Boni. The Nilotics arrived from the northwest via Uganda and Ethiopia, bringing farming skills, and settling around Lake Victoria and the rift valley, areas now dominated by their descendants, the Luo, Samburu, Maasai, Turkana, Nandi, Tugen, and Kipsigis. The Bantu group occupies almost half of Africa south of the Sahara, migrating from the west with roots in Cameroon. They brought farming, iron-working skills, and knowledge of root crop farming. Their descendants are the Kikuyu, Kamba, Meru, Embu, Luhya, and Mijikenda at the coast.

Around 1400 AD, Arabs and other explorers started arriving, with the seasonal monsoon winds providing a source of safe sea voyage in their sailboats called *dhows*. They would sail with their cargo east to the Indian sub-continent and, for its proximity, more frequently south to East Africa, settling along the East African coast. They introduced their culture, religion of Islam, fishing, craft industries, and trade flourishing to towns such as Pate, Lamu, Malindi, and Mombasa. The traders brought cloth, beads, and iron, which they exchanged for ivory, timber, gold, rhino horns, animal skins, and slaves. In the 1800s, slave trade dominated by Arab traders flourished in East Africa and Zanzibar was a thriving centre from which control was maintained. Slaves were brought here and held in gruesome, confined conditions before they were either sent to work on clove plantations or sold to clients in the Middle East.

Early trade in dhows during Monsoon season from and to the Persian Gulf and East Africa, 1800s.

With such dominance and commercial success, the sultan of Oman, Seyyid Said, proclaimed sovereignty over Zanzibar. In the mid 1850s as a result of conflict for succession, his brother, Majid Bin Said, assumed control of Zanzibar and moved his headquarters from Muscat to Stone Town and made it the capital of Zanzibar. With increasing trade on the East African coast and inland, he extended his sovereignty over a ten-mile coastal strip in Kenya.

From Europe, the Portuguese were the first to arrive on the East African coast in 1498 when the famous explorer Vasco da Gama stopped to refuel in Mombasa on his way to the East Indies.

His account of a thriving trade in the Indian Ocean with Mombasa as a trading hub was well received. The Portuguese started arriving in large numbers to trade and settle, making alliances with Arab sultans who ran the city states and were largely from Oman. The Portuguese used Mombasa as a trading hub for years. As the Portuguese influence grew, they constructed a fort in Mombasa, Fort Jesus, to defend the city state against invaders from the sea. To this day, this fort stands out as an architectural wonder of a bygone era, and has become a popular tourist attraction.

The Portuguese control in East Africa lasted 200 years, to about 1700, by which point local resistance to their dominance was increasing. An incident in 1631 was a turning point when the local Omani sultan of Mombasa, Don Chingulia, aided by Omani Arabs, attacked Fort Jesus. He killed the commander, the troops guarding the fort, and most Portuguese residents of Mombasa, destroying and burning their houses. The situation created panic, fear, and uncertainty for the future, and sent shock waves to the Portuguese.

The resistance by the local Swahili-Arab residents to foreign domination increased, and the situation became volatile. Acts of banditry worsened, forcing the Portuguese to begin to leave. By 1698, the Arabs conquered Mombasa and drove out all the Portuguese.

On the strength of the stories of Portuguese successes there, Eastern Africa, stretching from the Indian Ocean inland, was now attracting the interest of other European powers, including Britain, Germany, and France. British influence was gaining interest, with David Livingstone's and other missionaries' accounts about the wealth of the region and thriving trade in ivory and slaves.

Around the same time, Cecil Rhodes was signing up local chiefs in Southern Africa to bring them under British control.

He succeeded in establishing Southern Rhodesia, now Zimbabwe, and Northern Rhodesia, now Zambia. After the Boer War of 1899, Orange Free State and Transvaal in South Africa came under British control.

In 1886, European rivalry in Eastern Africa increased, with Britain and Germany vying for control. A comprehensive agreement was reached at the Berlin Conference to divide areas of influence from the Indian Ocean to the Great Lakes and respect each other's territories. An Anglo-German boundary commission, in consultation with France, was completed, which left a ten-mile strip on the Kenyan coast to the Sultan. The inland territories were divided. A straight line was drawn from Pemba on the Indian Ocean coast going north of Mount Kilimanjaro to Lake Victoria and the Nile watershed. Britain was allocated the area to the north and Germany got areas to the south. The line remains to this day as the boundary between Kenya and Tanganyika, now called Tanzania.

In 1888, a royal charter was given to the newly formed Imperial British East Africa Company (IBEAC) to acquire more territories, as the area was becoming important for trade. So as not to be landlocked, Britain, through IBEAC, extended its rights and control over the ten-mile coastal strip in consultation with the sultan of Zanzibar for a fifty-year lease, and called it a protectorate. The new country was renamed the Colony and Protectorate of Kenya. But the IBEAC did not have handy income and could not cope with increasing hostility from the Natives to their authority.

To make matters worse, in 1890, German Karl Peters forced Kabaka, the most powerful ruler of the Kingdom of Buganda, in Uganda, to cede his authority to Germany. European rivalry for dominance continued to increase in the region. Uganda became another area of tension. In early 1892, the Kingdom of Buganda witnessed instability at its capital, Kampala, which sits on four hills.

On one hill was the palace of the ruler, the Kabaka; the second hill had a British fort built by and named after Fredrick Lugard; the third had a Catholic cathedral built by the French; and the fourth hill had a cathedral built on it by the Protestants from Britain. Lugard prevailed, but he was not able to control and run the Kingdom of Buganda or the neighbouring states.

A crisis was averted when Britain ceded control of the tiny island of Heligoland, which had been in British possession in the early 1880s, in return for German recognition of the British influence in Zanzibar, Uganda, and Southern Sudan. In later years, Heligoland would become an important naval base for Germany in both the world wars. In 1896, Britain declared a protectorate over Uganda and continued to get more involved in the country. The rich farmlands from Uganda through the Kenyan highlands presented a unique opportunity to develop. A settler community was required to inhabit these lands and start farming operations for cash crops such as tea, coffee, wheat, sugar, and barley.

The road network was inadequate and subject to weather. During rains, roads became impassable as small streams became huge rivers. There were hardly any bridges, and motorists had to wait until water subsided. This could take days and, even then, driving conditions on soggy, soft, muddy roads meant getting bogged down for even longer. The British recognized a need for a railway line to stretch from the Indian Ocean port of Mombasa inland to Kampala on Lake Victoria. This would encourage the newly arriving white-settler community to grow cash crops with the assurance that, after harvest, they would be able to transport their produce to the rest of the country and for export via the Port of Mombasa.

Economic activity had to be created to generate revenue to pay for running the colonies and provide security for the newly arriving settler community. Britain financed the construction of a railway to open up the rich interior farmlands which were suitable for coffee, tea, tobacco, and other cash crops. They brought in indentured labour from India with skills to construct the railway, which was completed in 1906. Some 22,000 workers were brought from India for manual labour.

The first local grievance was from the Maasai, who were relocated by the British from the fertile Laikipia district to arid Ngong. The Maasai felt that they had been deprived of their inheritance. Local native councils were formed with an elected chief supported by a younger educated youth. The Nandi rebellion against British rule was promoted by a Nandi leader, Koitalel Arap Samoei, who spread the word that a "black snake"—the railway line—would tear their region apart, spitting fumes.

Resistance to British rule intensified. Slave trading was brisk and the Arab traders felt threatened that their monopoly was being interfered with. They carried out raids and ambushes to frustrate the aims and influence of IBEAC, which could not handle this worsening situation as it did not have the finances or armed forces to defend. In 1895, IBEAC sold its interests to the British government.

Britain continued to expand its influence in East Africa, from the Indian Ocean inland up to the Great Lakes, encompassing Kenya, Uganda, Tanganyika, and Northern and Southern Rhodesia.

First, the missionaries started arriving to spread Christianity. This was followed by the imperial powers, whose interests were commercial and in pursuit of expanding their influence. The church was pursuing its own interests by sending missionaries inland to disseminate its word. This was achieving much success, and new converts were gradually absorbed into medical clinics, schools, and the churches that were slowly being set up in remote areas. The missionaries' goals included the elimination of slavery, a development that brought them into conflict with the slave traders. Commercial activity on the island grew, and the sultan brought in Indian migrants to develop the clove plantations with slave labour. Cloves became a mainstay of trade. With pressure from the British, the sultan reluctantly agreed to stop slave trade in 1873.

In 1900, Britain created the East Africa Common Services Commission with Kenya and Uganda so that, federally, the countries could be well managed. During WW I, Kenya was an actual theatre of war. In 1914, Colonel Paul von Lettow-Vorbeck, commander of German troops in Tanganyika, waged a highly successful guerilla war to capture British supplies and remained undefeated. He trained local soldiers, *askaris,* to fight the British. Britain used Indian troops under the South African commander, General Jan Smuts, and kept the Kings African Rifles (KAR) on internal security operations.

Britain was coping with security on the newly constructed railway to Uganda and fighting German troops in Tanganyika who regularly attacked trains for supplies. During WW I, Britain raised regiments of porters to supply the interior by foot. They were called 'Carrier Corps' and they mobilized over 500,000 natives. A large number of foreign troops in the interior had to be fed and supplied over long distances without convenience of rail or road. To deliver one kg of rice, the carrier corps started with 50kg at the coast, the balance being for use by the carrier corps. They were also used to sustain logistical supplies to those British troops who were engaged in anti-cattle-rustling operations, internal security duties, and fighting German troops in Tanganyika. Carrier Corps was a military organization with a proper chain of command. They had camps outside major towns and cities where the men were barracked. These towns acquired the name *Kariokor*—a derivative of the name Carrier Corps—which exists to this day. In 1916 and at the end of WW I with the Treaty of Versailles, Germany gave up claims to all its territories in Africa.

World War II had a great influence on Africans' thinking. They did not understand why they had to leave their families and endure suffering and hardship to fight a foreign war about which they knew nothing, a war that had them fighting white soldiers and killing them with apparent ease. This reinforced the notion in their minds that they could "kill the white man" at home.

DECOLONIZATION

The colonies in Africa started agitating for independence early in the twentieth century. Various movements wanted a voice in running their affairs. Their aim was to reclaim the lands that had been forcibly taken from them by the European powers during their quest to build empires.

The only countries not colonized were Liberia on the west coast, which was settled by freed slaves from the United States, and Ethiopia, which resisted a brief Italian occupation just before WW II.

Agitation against colonialism continued far and beyond the rural villages and countryside in Kenya. The Kikuyu were the most aggrieved, and they wanted their land back, which had been taken away from them forcibly. Despite being labelled treasonable with consequences of prosecution, many spoke openly in public and private gatherings to reclaim their land.

The stand taken by a soldier of the King's African Rifles, Waruhiu Itote, conveys a universal simmering discontent.

As a soldier during WW II in the Kings African Rifles, he served bravely in Ceylon (present-day Sri Lanka) and Burma. He was proud that he was promoted corporal whilst in Burma for his extraordinary skills as a tracker, mastery of field craft, and natural friendly personality for leading and producing results. He would recall many instances with immense pride of proudly serving in Field Marshal William Slim's 14th Army in Burma. He fondly remembered what a great leader Slim was for the simple reason that, despite being the senior-most officer, he still found time to visit troops in camps. On such visits, he gave pep talks to raise spirits, instill morale, and justify why the Japanese had to be driven out.

On his return from Burma, his outlook changed. He saw that his family had been driven out of their traditional farm holding by the colonialists to the point of begging. This infuriated Itote. He could not reconcile how, on the one hand, he went to Burma to defend *Kingi Georgi's* empire, risking his life in a far-off land that he knew nothing about, but at home, these very colonialists were driving his family to the brink of extreme hardship and poverty as a policy. The squatters issue, the *kipande* issue, the hut and poll tax issue, and a general slavery kind of society added to his unhappiness and made him bitter.

Itote decided that a bigger cause lay ahead and that he would now make it his life's goal to drive out the very colonialists whom he had loyally served. He resigned from the Kings African Rifles, joined the Mau Mau movement, and took refuge in the Mount Kenya region. He was a close confidant of Jomo Kenyatta, but would not elaborate on his relationship other than reaffirming that Kenyatta was the leader of the Kikuyu. With Itote's natural leadership qualities and ability to organize a body of men, qualities that were recognized and acknowledged with a field promotion in Burma, he quickly rose within the Mau Mau ranks. He was appointed by his peers and elders to command the movement in the Mount Kenya region from where he would direct operations against the colonialists. These operations took the form of laying ambushes and intimidating other Kikuyu loyalists who worked for the settlers to join the Mau Mau under threats of beheading.

Mau Mau had no heavy weaponry; their tactics were to hit soft targets at night—often the loyal Kikuyu who did not submit to the Mau Mau cause. With a decentralized command structure, a Kikuyu was able to operate alone, if needed. Itote's numerous operational successes, loyal following, and swelling recruitment earned him a variety of pseudonoms. General China is the name he liked and adopted.

Waruhiu Itote, by now widely known as "General China," played a leading role in directing the movement in the Mount Kenya area. Others were Dedan Kimathi, who operated in the Aberdares Mountains, and Stanley Mathenge, in Central Kenya.

He was captured in 1954 on the way to a meeting in the dense Mount Kenya forest by the legendary British spymaster, Ian Henderson. Itote was sentenced to death by hanging for his membership and role in a banned and illegal society.

Recognizing the potential to use Itote to encourage his fellow combatants to surrender, Henderson was able to negotiate for his sentence to be commuted to life in prison. Itote was dispatched to remote Lokitaung, where he spent five years in isolation. He was later joined by several vocal leaders who were vigorously promoting independence, such as Jomo Kenyatta, Achieng Oneko, Bildad Kaggia, Fred Kubai, Kungu Karumba, and Paul Ngei.

In post-independent Kenya, I met Waruhiu Itote many times and we developed a warm, friendly relationship. He was appointed director of the National Youth Service, a government agency tasked with training a young growing population with skills and trades for the job market.

After I had moved to Canada and on my numerous visits to Kenya, I would call on him. He was always gracious, welcoming, and hospitable. He would invite me home for a traditional *irio* meal, and would joke: "In Canada, there are no Kikuyus and you better have it here as you will not get it there!" We would discuss the future of the country, the widening gap between tribes, and the increasing disparity between a growing rich native upper class and the volatile, large, restless, and landless segment of society. He rarely discussed his role in the Mau Mau movement and, whenever the subject came into a discussion, he would direct me to history books!

The decolonization process speeded after WW II with US President Franklin Roosevelt and British Prime Minister Winston Churchill signing the Atlantic Charter in 1941, under which Britain agreed to expedite the granting of independence to the colonies.

In 1947, Britain granted independence to India and other European powers also granted independence to their colonies. World War II actually sped the decolonization process, since the two emerging superpowers, the United States and the Soviet Union, were anti-colonial in their outlook and urged independence movements in their sphere to continue to agitate in that direction. As the Cold War was splitting countries into their spheres of influence, Britain guided as much of its former empire into the British Commonwealth as it could, promoting democracy, free enterprise, and a leaning toward the West and the United States. Those countries not wishing to align with either superpower created a non-aligned bloc, the Non-Aligned Movement, at the Bandung Conference in Indonesia, in 1955.

In the early 1950s, Britain created a federation of Kenya, Uganda, and Tanganyika, called the East African Common Services Organization, to unify common services as a measure for economy and efficiency. This included East African airways, railways, ports and harbours, posts and telecommunications, inland revenue, meteorological services, and defence. There was free movement of employees of these agencies among the three countries, and these institutions handled their tasks well and to efficient standards.

In Kenya, various independence movements began agitating to regain their lands early in the 1900s. These activities culminated in the Mau Mau rebellion in 1952.

Mau Mau was a nationalistic movement to drive the Europeans from Kenya who had illegitimately confiscated rich farm lands in the central part of the country and driven the traditional dwellers, the Kikuyu, away.

The displaced Kikuyu became restless and, with limited choices, took up labour and low-paid employment on the very farms they once owned. Others went to the cities in search of low-level clerical government jobs or domestic positions. The Kikuyu highlands, also referred to as the white highlands, stretch from Mount Kenya to parts of the Great Rift Valley, the Aberdares, measuring some 30,000 square kilometres in close proximity to the capital, Nairobi. Among Kenya's most fertile farmlands, they were forcibly confiscated by the British in the early 1900s in keeping with Britain's quest to build and sustain the empire. At over 5,000 feet in altitude, this part of the country experiences warm days and cool nights. It is mosquito-free and, hence, free of malaria.

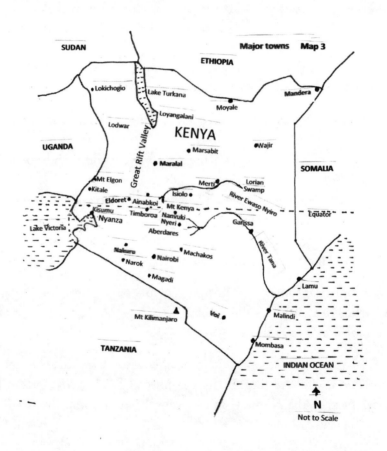

Major towns, Kenya.

PRINCESS ELIZABETH BECOMES QUEEN

The year 1952 is a historical one for the British monarchy, and also for Nyeri, a small town in the heartland of Kikuyu country, which has earned a place in history for the monarchy, the Mau Mau, and the world scouting movement. A young Princess Elizabeth was holidaying with her husband, Prince Phillip, and an entourage of royalists, including Lady Pamela Mountbatten, wife of Lord Mountbatten. They were in Kenya as part of her tour to Australia and New Zealand, which started on January 31, 1952. They were seen off in London by an ailing King George VI. Some have maintained that the timing of her visit to Kenya was to reinforce confidence in the settler community of London's continuing support.

On February 6, 1952, they were staying at the Treetops Hotel in Nyeri, when the princess's father, King George VI, died in London of lung cancer. Winston Churchill broke the news to the world from his residence at 10 Downing St.

The Treetops Hotel has a history of its own. It was built as a game-viewing lodge on a sacred Mugumo tree, overlooking an elephant watering hole where many wild animals such as leopards, rhinos, and wildebeest came for water.

This was the only game lodge of its kind in the world at that time. Jim Corbett, a naturalist and conservationist who had retired from his conservationist work in India and was accompanying the royal couple described the development thus: "She went up a tree in Africa as a princess and came down a queen."

Adding to its significance was the parent hotel of the Treetops, the Outspan Hotel, which was built in Nyeri by Eric Walker, an ex-RAF pilot who was shot down over Germany during World War I. He escaped to Canada and started an illegal liquor export business during prohibition. When the authorities went after him, he took off, migrated to Kenya, and married Lady Bettie, daughter of the Earl of Denbigh. He thus secured a place in royalty with its trappings of privilege and upward social mobility. With such credentials, he was designated a host for the royal couple. Corbett lived in a cottage at the Outspan that was once occupied by Lord Baden-Powell, founder of the world scouting movement that he started in South Africa. There are stories that Baden-Powell had smuggled a pair of pliers in for Eric Walker when he was a prisoner of war in a German prison camp during WW II. Walker used these to cut out an opening in the barbed wire enclosure to escape. On migrating to Kenya, Walker became private secretary to Baden-Powell, who lived in Nyeri and is buried there. His grave has since become a landmark, and enjoys a place on the tourist calendar.

In the early 1960s, East Africa witnessed turmoil in the region as Britain granted independence to Tanganyika in 1962, and to Zanzibar in 1964. Both countries became self-governing. Shortly thereafter, a revolution took place in Zanzibar, with the People's Republic of Zanzibar and Pemba, a neighbouring island, declaring independence. In 1964, the island was united with Tanganyika. It retained self autonomy with the new country, now named Tanzania. Kenya had a mild turmoil in 1964 when a small number of troops from the 11th Battalion KAR mutinied at Lanet. This was put down immediately.

MIGRATING FROM INDIA TO KENYA

In the early 1900s, with Britain ruling India, wide publicity attracted volunteers to emigrate to far-off lands within the British Empire. They were given promises of well-paying jobs, housing, generous benefits, and, most importantly, pension benefits, the most sought-after perk at retirement. From 1895 onward, a large migration took place from India to Eastern Africa. These skilled emigrants were brought in as indentured labourers to work on building the Mombasa-to-Kampala railway and opening the interior. As word got around that Africa presented huge economic opportunities, more Indians—primarily from the agriculture-rich Punjab and business-inclined Gujarat regions—started migrating in large numbers that resembled a kind of gold rush. The Indian population was primarily of Hindu, Sikh, Muslim, and Christian faith, and they brought in their religious elders since the social structure was tightly knit and centred around their faith. The *pandits* for the Hindus, *mullahs* for the Muslims, priests for the Christians, and *gyanis* for the Sikhs followed. On arrival in Kenya, they set up places of worship wherever they could find space: in houses, tented camps, and schools. From here, they conducted weekly services and provided a resource for those seeking spiritual support.

Father brought Mother from the village of Kartarpur in the Punjab in 1942. After a few days staying with friends in Mombasa and Nairobi, they moved to the interior. I was born at home, delivered by a midwife in Ainabkoi, a little-known village in the hilly forested central Kenyan highlands. The nearest town was Timboroa, which claimed the honour of having a railway station at one of the highest altitudes—4,000 metres—in the British Empire. Ainabkoi was a small trading centre with a population of about 100. Half of the residents ran small shops called *dukas* and the remainder worked for the government, mainly in a health clinic, public works, the police, and forest department. The nearest hospital was over 100 kilometres away in another small town, Eldoret. Roads were made of *murram*, a hard soil compacted with crushed gravel. In dry weather, vehicles would spew out dust and keeping 100 metres behind a car or truck was normal. During rainy season, the low-lying parts of the road would be impassable with dry river beds becoming torrents. Bridges were few, made of simple wooden supports over which wooden planks rested. There were no asphalt roads, no electricity, and—as such—no street lighting.

Only the medical clinic and the police station had kerosene-powered generators, which were used sparingly to conserve fuel.

Ainabkoi: My early life

My early life in the interior of Africa was devoid of material possessions. It was spartan and included many of the same challenges that the early pioneers had endured. With limited access to affluent material possessions, life was a journey of survival in the hope that better days lay ahead. Given that our folks had left the Indian subcontinent in search of better opportunities, we had to make the most with the least. There was no going back and, as a first generation born in Kenya, this was home. Overcoming numerous obstacles and challenges was part and parcel of life. We faced regular choices between need and want and quickly learned how to manage our limited resources with creativity and improvisation. Such a routine made us appreciate what we had, strengthened our characters, and matured us at an early age.

The majority of newly arrived Indians were single men. They left wives and children back in their villages and took up whatever job was available without being selective. Most of these jobs were in the embryonic civil service or skeletal government agencies such as the railways and the post office. Some took up trading by opening small shops or kiosks.

One such trader launched his beginnings by buying a ten-stick pack of Sportsman cigarettes for fifty cents. He would stand on street corners with a tray slung around his neck selling these sticks at ten cents apiece—easily making a 100% profit.

Another entrepreneur would buy a twenty-four-inch bar of Sunlight or Lifebuoy soap, popular in that era, for one shilling, and sell cut-out twelve-two-inch pieces for twenty cents each, a 140% profit! Still another would sit on a wooden box on a busy street corner to repair punctured bicycle tubes and straighten wheels for a worldly sum of ten cents. The common mode of transportation was bicycle, and Raleigh was a popular brand. The bumpy, potholed roads meant bicycle repair was a thriving sector.

With ingenuity and in time, these gentlemen rose to prominence by developing huge industrial conglomerates. They would employ a large number of people, give them skills and trades, and thus contribute to the national economy (and, in the course, make a lot of money).

My father was a surveyor, a profession much in need, and he took a secure job with the forest department. The job expected a wide range of skills, some of which were acquired on site, and was located in the remotest part of the country. His routine was of a nomadic type, requiring numerous trips into the interior forests to log plant species, stake out sections for logging and road construction, and identify locations where trading centres could be established. Such work required having a caravan of porters, a cook, horses, and a large number of assistants and guards accompany him wherever he went. He would set up a temporary tented camp and carry field camp beds, sleeping bags, lanterns, and rations of eggs, potatoes, vegetables, and rice for the team. For meat, they carried live chickens that were slaughtered on site. Milk, if and when available, was often procured by team members from their clansmen or locals in the vicinity. Water was simply drawn from nearby streams and there was never a thought of purifying it as that is how the locals drank it without fear of disease. One's natural immunity to all kinds of exposure other than malaria was superb. The campsite would be set up for three or four days, and then moved as his field work progressed.

A medical orderly accompanied them, carrying a basic first aid kit, but since he had a good knowledge of indigenous plants, he resorted to administering traditional field concoctions and they worked. For the period he was away, there was no communication with his office. For a rare emergency, a runner would be sent on horseback with a message. Guards were required to ward off wild animals and reception from the natives was warm and friendly as they did not see him as a threat. They were more curious about the intrusion into their hinterland territory where no foreigners had ventured before. If he went away for one or two days, he would take me with him; for longer trips, he would go alone.

This was the beginning of my love for the outdoors and a deep understanding of the Natives, which happily formed part of my personality and conditioned my outlook on life. Much greater rewards awaited in later years.

The biggest benefit of his demanding work routine was a three-month vacation back to India, taken every three years. Travelling to and from the Indian subcontinent was undertaken by boat, and the popular steamships were the SS Karanja, SS Kampala, and SS The State of Bombay. Travel would be third class by choice and folks would carry their two-week supply of snacks to complement the ship's self-serve buffet meals. They also carried holdalls—cotton-filled quilts and pillows that were used for sleeping. This option was preferred as the difference in price of a second-class ticket and a deck one was significant. Plus, the camaraderie on deck meant plenty of entertainment—swapping stories and watching movies every evening—and the time would just pass by.

Travel back to the ancestral village of Hoshiarpur in the Punjab was a journey to catch up with news of family births and deaths, and to hear of those who had migrated to far-off lands in search of opportunities and a better life.

An arrival in the village resembled a hero's welcome, with relatives and neighbours gathering around to hear stories of far-off lands and see what gifts or money was brought for them. They were curious to find out how life fared, whether food was plentifully available, what medical facilities provided care, how he overcame language barriers, and how the locals welcomed him.

Father would give whatever little money he had brought to his relatives as a way of supporting them. Such support was often to fix a leaking roof, repair a door, or pay off debts at the local grocery store.

Self-esteem and pride to uphold the family name meant that having a financial debt was fine, but it had to be paid off to avoid being labelled untrustworthy. Village life was very close and everyone knew each other. Neighbours played the role of extended family, exercising authority over others without fear of trespassing into their space.

A family rarely enjoyed privacy since the clan's and neighbourhood's interests were above the family's. Being secretive amounted to being mischievous.

The village life had its norms, for customs and rituals steeped in tradition were not to be tempered with. Every family would have a priest, a *pandit,* who played an important role in their life, exercised enormous power and influence, and was worshipped like god. Superstition and fear ruled the day. The *pandit* would be consulted prior to arranging a date and timing for any family event, from birth to death. The *pandit* would consult his astrological data and, after computing the alignment of the stars, recommend the most suitable date and time for an event such as a wedding.

It was quite normal and widely accepted if, for example, the timing for a wedding was two o'clock in the morning! The *pandit* was never to be questioned, since questioning him amounted to challenging his authority. Even naming a newborn was determined by the *pandit.*

Parents of a newborn would consult the *pandit* who would give two or three alphabets upon which the parents would submit a list of names starting with the recommended alphabet and, on his advice, a lucky name would be selected. The *pandit's* ruling on any issue was final. Social standing in the village was predetermined, and it was of utmost importance to uphold the family's honour.

Age had an automatic status of seniority and demanded submission and total respect. The elders would automatically form part of the village court, the *panchayat,* and would handle minor conflicts and disputes and pass judgement arbitrarily with the expectation of total compliance without recourse to an appeal. Such was the system and it was well-entrenched. It worked and resulted in village stability. I was to find later that such a clan or tribal system also existed in the African society.

Our house was a single-story, wood-framed structure with a corrugated iron-sheet roof. We had no plumbing, no running water, no electricity, and no telephone. We had to collect mail from counter clerks at the local post office once a week—or, more likely, once a month, since letter writing within Kenya was a rarity.

Our first house in Ainabkoi, with water storage tank in front as a standard feature, 1948

The only anticipation was receiving a letter via surface mail from India, which would take four to six weeks. Interest in local news was primarily for community services and to find out who else had migrated from the subcontinent, especially those from one's own village or district.

The rallying point for exchanging news and local gossip was the local barber's shop. Here, men would gather for tea after work and to swap news and stories.

The barber's shop doubled as a social centre. When Father would get home, we would be entertained to news: who else had migrated and how they were progressing in their job search, births and deaths, upcoming weddings, and community celebrations in places of worship, be they in churches, temples, *gurdwaras*, or mosques. No invitation, written or verbal, was necessary, and since a celebration was held in place of worship, it signified a communal social event. Everyone available was expected to attend and partake in cooking and serving food. No alcohol was served as it was prohibitively expensive—beyond affordability—and, besides, its consumption was looked down upon.

There was no animosity within a new mushrooming society based on religion, caste, or origin; the social goal was to be respectful and supportive of each other since all the newcomers had similar backgrounds, resources, and aspirations for a promising future. We all respected each other's religion with its values and restrictions as a matter of social accommodation.

In later years as the community grew, schools were segregated on religious grounds and a secondary language of instruction became mandatory. With different alphabets and written work specific to that religion, the government regulated a system that separated people, thus leading to social divisions. The main language of instruction was English. Although Swahili was the native language, it was not taught in schools. Whatever little Swahili we understood and spoke, we learned at home.

The most reliable method of communication was by sending and receiving a cable through the post office, relayed by Morse code. A cable was only sent or received when someone died or, rarely, to announce a birth. Letters were exchanged and the preferred mode was surface mail which took, on average, a month from Kenya to India.

At home, we had a fireplace running daily to ward off the highlands chill. Abundant firewood was available from the nearby forest, and the fireplace was stoked every night. Cooking was done on a firewood-stoked Franklin stove, which also provided heat for the house. Mother would often place washed clothes on racks by the stove to dry if the outside was wet or dusty.

We had four domestic hands whose duties were clear. Njoroge had the task of ensuring we had water. He would run daily errands to fetch water in a cart, pulled by a bullock with a forty-gallon recycled oil drum in its carriage that had been laboriously cleaned. He would leave home at sunrise, about 6 a.m., fill up the drum manually using a bucket at the local stream some five kilometres away, and return by midday. Such was his daily routine. This was pure, clean water manually taken from a running stream that we would drink happily without fear of contracting any waterborne diseases. For baths, we used a bucket and mug filled with water that was heated using firewood as fuel in four-gallon recycled and thoroughly cleaned oil cans. We also had a huge 800-gallon water-storage tank for capturing rainwater from the roof—a standard feature of houses. An elaborate system of eavestroughs and drain pipes directed water into the tank reservoir. I can't recall ever using any water-purification additives.

Daily water delivery from a stream 5 km away.

The second domestic hand was Wambui, a nanny called *ayah* who would take care of us children. The third domestic hand was Mwangi, the servant who took care of house-cleaning and the garden.

The fourth, Kiplangat, was a guard equipped with a bow and arrow whose job was to ensure that no wild animals came close. All four domestics were accommodated in simple one-room quarters, about fifty metres from the house.

Crime of any nature was not heard of and, when going out, we never locked our doors. At night, we would secure the door with a wooden bar placed across brackets.

The early toilets were simply outhouses some twenty metres downwind with a deep hole in the ground, on top of which was a wooden platform to which a toilet seat was affixed. Privacy was achieved by a simple burlap wrap around a fence supported by wooden poles.

Later, as other residents moved into the neighbourhood, these toilet enclosures were improved. Looking much like sentry boxes, they were upgraded with a high seat and bucket into which one pooed. A small towel left by the entrance signified that someone was using the toilet. Water was used to clean oneself. Toilet buckets were emptied every night by local contractors called *dias* who would do their work—dumping the contents in a pit in the forest—discreetly, as the profession was looked down upon.

In later years almost up to the mid 1950s when municipal sewage and water services were introduced, such outhouse toilet facilities were a common feature in houses in Nairobi.

We had no electricity. For going outside at night, we would use a kerosene-operated hurricane lamp or flashlight. Lighting inside was with hurricane lamps, pressure lamps, and candles made of locally collected beeswax. Any homework that could not be completed in daylight was done by the candlelight or hurricane lamp. There was no television. The only external entertainment was listening to the one or two channels that were audible with a battery-powered radio with poor reception. To conserve precious and expensive batteries, the radio was switched on for half an hour every evening at dinner time to hear world news from the BBC, which was relayed via a local radio station.

We had a manually cranked gramophone, His Master's Voice, with a bugle-type audio speaker that used 33 rpm vinyl records. When we got a kerosene-powered generator in the early 1950s, this signified a major achievement and became a talk of the town.

Telephones were a luxury. We had one installed around 1950. It had a crank handle. To make a call, we would crank it several times to speak to the local operator who would manually connect us to our desired number. Or we would book a call with a twenty-four-hour notice and wait patiently around the phone for it to come through. A booked call had to be timed: two or three minutes and we would be interrupted by the operator at the thirty-seconds-remaining point. With such an expensive and time-consuming procedure, we quickly learned to be brief and to the point on phone, mentioning only the main items without going into unnecessary pleasantries or detailed discussions. A phone line was also shared by others in the neighbourhood, who could jump into a call at any time or overhear our conversation without any concerns about being nosy or inquisitive. Gossiping and conversations about the neighbours were therefore out. Phone numbers were in single digits in small towns and double digits in bigger towns, as phones were a scarcity. Subscribers in Nairobi, with a population of 100,000 in 1950, had three-digit numbers.

Parental roles were clearly defined. The husband was the breadwinner in the workforce and the wife stayed at home looking after the children, cooking, sewing, and attending to any family-related issues. Father was the authority.

In Ainabkoi, we were one of about ten families with school-going children ranging in age from toddlers to twelve. Of these, three were Brits who pretty well kept to themselves in their small private clubs. These were hidden away in exclusive preserves where weekend tennis or bowling were the norm, enjoyed over pinkies (gin made with Worcestershire sauce that gave it a pink colour), and followed by a curry lunch. Their contact with newly arrived immigrants, most of whom were from the Indian subcontinent, was only at work since socializing with WOGS— Westernized Oriental Gentlemen—was looked down upon by peers.

We had no school but rudimentary classes were held under a shady tree with us sitting on logs listening to rotating volunteers who acted as schoolteachers, imparting whatever knowledge they had from their younger days. These volunteers were civil servants or shopkeepers. We were taught collectively, regardless of the subject matter, but the focus was on mathematics, geography, and English literature.

Our writing pads were wooden boards much like cheese boards, and we would use dry, angle-cut reed pens, picked by the river bed and dipped into an ink well to write. This archaic method was eventually replaced with a 6"-by-12" slate with a wooden frame and chalk, and then by proper paper and a holder with a nib that we would dip into an inkwell to write. Our parents' and neighbours' sincere goals in such stark adversity were noble and futuristic: to instil in us a sense of fulfilling daily routines and not just playing around. We had to get a solid foundation and a firm base for an education in the future when the opportunity arose to attend a proper school.

All of this was conditional on Father being posted to a place that had a school in close proximity. Options for boarding school were restricted to children of British and European families who had the financial means and access; non-whites were not allowed to attend. Race had its perimeters that were not to be crossed. Words such as whites, Indians, coloureds, natives, blacks, and Africans were part of a standard vocabulary that was used commonly to identify racial lineage. Racial segregation at the workplace, in homes, and in public transportation was rigorously enforced.

My childhood till age twelve was spent moving around small townships: Londiani, Mau Summit, Kedowa, Kaptaget, Elburgon, Plateau, Timboroa, and Eldoret. We stayed at a place for a year or less, and my schooling was erratic and uncertain. English was the main language of instruction, but we had to have a second language of our ethnicity—but not the native language of Kiswahili. The second language would be one of the main languages of the Indian subcontinent as represented by the new immigrants: Hindi, Punjabi, Gujarati, or Urdu.

Since small schools catered for a second language only if there were sufficient numbers, I took Urdu in grade one at age six, Gujarati in grade six Punjabi in grade seven, and Hindi in grade eight. Father sent me to live with close family friends in Nairobi to attend grades nine and ten, as these grades were not offered in the very few schools in rural Kenya. Perhaps it was because of switching several languages without a choice at an early age that I developed an ease with learning new languages.

As I was approaching high-school age, Father requested a transfer to manage Karura Forest, on the outskirts of Nairobi.

MOVE TO NAIROBI

We all moved to Nairobi in 1954, settling into a government-supplied brick-and-mortar house. We got stability for the first time, with me continuing onto high school at the Duke of Gloucester School. Schooling was segregated, with distinct barriers between communities whose sole aim was to keep the growing, hardworking, and industrious community separate, divided, and, at times, adversarial.

This was a relic of the British divide and rule policies practised in every colony. They were enforced to keep communities apart and stop them from becoming cohesive out of fear that their members would start demanding rights and privileges.

The Europeans, Indians, and Natives attended separate schools in accordance with their race and ethnicity, a rule that was vigorously enforced by not even staging inter-school activities such as sports or debates. Schools were further divided according to faith. As such, Hindus, Sikhs, and Muslims attended schools of their faith, while Goans attended Christian schools. Goans hailed from the former Portuguese-controlled Daman, Diu, and Goa enclaves on the Western Indian coast. They attended Christian schools since the Portuguese forcibly introduced Christianity and vigorously spread it by converting Hindus to Catholicism. No non-white teachers taught at European schools, though white teachers would teach at Indian and African schools.

A European was the principal of every school and the teachers were from the race and ethnicity of whichever school they taught in. In keeping with the colonialists' divide-and-rule policies, residential areas were also separated by race, and no one was allowed to live in any area other than those designated by their ethnicity.

The only ones allowed to live wherever they chose were native male Africans. By no other choice, they would take employment as domestic hands, commonly referred to as *servants,* and were housed in outhouses with minimal basic comforts. They, too, were discouraged from bringing their wives and children as a policy, though they were allowed to visit their families in villages every three to six months for two or three days. Some women were hired as *ayahas,* and had to take employment without their husbands' accompaniment. They were also allowed to visit their families in their village every three or four months.

Clothes were handed down, or Mother would sew them at home using a Singer or Pfaff sewing machine—the most extensively used item in any household. Worn-out shirt collars would be turned inside out, and shoes would be resoled by the town cobbler, who enjoyed an important social status along with the town barber. There was no blatant consumerism or brand-label affiliation. Cotton and wool were the materials of choice, as polyester or other synthetics were still being developed. Children up to the age of about fifteen wore shorts, as trousers were meant for adults. Clothes were pressed with a heavy iron filled with charcoal.

Milk was delivered by a local farmer in a milk can that was placed at the end of the driveway every second day. Mother would produce yogurt, cream, and butter with it. Our meals were prepared fresh daily since we, like other families, did not have a refrigerator. As such, rarely was there any leftover food. Vegetables were home-grown in a patch behind the house called a *shamba.* Often, when socializing with other community members on a weekend or in some communal celebration, we would exchange our garden produce. Wheat and corn kernels would be taken to a local mill for grinding, and this would be the flour that Mother would use to make *chapattis* and bake bread.

Meals were what Mother prepared for all, and we happily ate whatever was on the table. We had three regular meals a day, all at home.

Going out to a café or restaurant were images only seen in movies, as there were none of these luxuries in small towns. Even Nairobi had very few restaurants. Eating out was almost the preserve of the white settlers who enjoyed the immense privilege of living their lives as they had lived in Britain or another European country. These few restaurants were located in a small number of hotels and were restricted to white settlers.

No Indians or Natives were allowed in unless they worked in the kitchen or bar, or as a member of the wait staff. Nairobi and the smaller up-country towns had sports clubs that became the venue for socializing and playing tennis, cricket, and bowling for the settlers. These sports clubs were also restricted to whites only. Non-whites could not apply for membership and it was extremely rare for a non-white to be invited as a guest to attend a function. Racial discrimination was enforced to the letter.

Groceries were the basics only, purchased at specialty stores by weight measured in pounds and ounces, using two plate scales with weights on one scale and goods on the other. Purchases were packed in cone-shaped newspaper wrapping. All transactions were in cash as the banking system was in its infancy and out of reach for most.

Shopkeepers maintained a credit system wherein all purchases were recorded in longhand in a ledger. The customer would settle his account at month's end when he received his pay, which was often in cash. The post office provided basic, rudimentary banking services where one would have a savings account, maintained by a bank book that was used to deposit and withdraw cash. In later years, travellers' cheques—Thompsons—appeared, and these offered a secure method to exchange funds in rupees on travels to India. Other travels requiring financial support were for a privileged few, primarily to Britain to attend a university.

All our emergency medical needs were stored at home. Our medicine cabinet had tincture of iodine, dressings, Band Aids, and quinine tablets. For minor issues such as cough and fevers, Mother would prepare homemade medications using ingredients from the kitchen spice containers—and they worked! A cold and cough remedy was made using a spoonful of turmeric powder mixed with milk and honey. For stomach upsets, we would be given granular seeds of anthem or fennel seeds, a spoonful of which we would keep in mouth and slowly suck. Fennel could also be mixed with bicarbonate soda powder and honey. Saffron was also used to cure coughs, stomach gas, and insomnia. Fenugreek was used externally for skin inflammations, ulcers, boils, and eczema.

For brushing teeth, we would use a twig of a widely available and special bush called *maswagi* that we picked in our backyard. It was two centimetres in diameter and six inches long. We would chew one end to make it bushy and brush our teeth. After brushing, we would gargle, rinse our mouth with water, and dispose of the twig. The branch contained chlorophyll, the ingredient that eventually found its way into factory-produced toothpaste. Using such natural, readily available, disposable products made our teeth strong and white, and I never heard of any tooth fillings or cavities. Even the profession of a dentist was distant. We never had a family dentist and there was no such thing as periodic dental check-ups. The nearest we got to a dentist happened when old folks had dentures made when they lost their teeth.

Malaria was to be kept away and, if Mother suspected signs of fever and a slight yellowing of face, she would give quinine tablets with water, two times a day, and that would be sufficient to cure. We did not stock or use any health supplements, as home-cooked meals were nutritious. We had healthy eyesight and wearing glasses, whether on a regular basis or for reading, was rare. Wearing glasses was a status symbol of academic achievement and confined to senior teachers. Also, wearing glasses all day was indicative of poor vision and, until the early 1960s, disqualified candidates for many jobs, including the army and the police force.

When a medication was deemed necessary, the local doctor would write a prescription and this would be taken to the local pharmacist, commonly referred to as a compounder. He would use a mortar to compound ingredients into powder to be taken with water. In time, the compounder profession evolved into that of a pharmacist.

We did not generate any garbage, and the word itself was not part of our daily vocabulary. There were no canned food products nor any disposable containers, jars, or bottles. Packaging was almost non-existent. The only packaging was a tall glass bottle of Sunglora sugar-laden orange juice concentrate, which we would dilute with water to drink. These bottles were recycled to carry or store milk. Slowly, jams and supplements such as Keppler cod liver oil started appearing and, after use, the containers would be cleaned and used to store spices and condiments. Municipal-issued metal garbage bins with lids were first introduced in the late 1950s. They were of such high quality that we took them in our homes for storing flour, corn, and sugar. With a large number disappearing, the city came up with a novel idea and replaced them with perforated holes at the bottom and on the sides to discourage their use as storage bins.

Cooking oil was purchased in four-gallon tin containers that, after use, would be thoroughly cleaned and taken to a blacksmith to have a lockable lid with a lip installed so they could be used to store flour, rice, and sugar.

Soft drinks of the Coca-Cola brand were available in half-pint glass bottles with a metal cap. After use, the empties were returned to the store for refund. A local newspaper—a two-to-four-page tabloid—was given back to the store after reading for use as packaging material.

The social make-up had stability since the authority of adults everywhere was respected as an extension of parents' roles. The school-teacher was like a god, never to be questioned as that would imply challenging authority. His ruling on any issue was final. Age and status had a standing and were respected without reservation. Any elder was addressed as *uncle* or *aunty*—addressing an elder by first name was not heard of.

The schoolteacher was always addressed as "Sir," since there were hardly any female teachers. Other community leaders who enjoyed a superior status were the local doctor, the policeman, the bus driver, the shopkeeper, and the religious head. Whenever a breach of any etiquette or mannerism came to the notice of parents, we would be scolded with the reasoning being that we were bringing shame to the family. Parents would side with whomever had reported. Corporal punishment at school for even the mildest infraction was routine, and was administered either by the principal or a teacher. This would take the form of a beating with a yardstick smacked on the hand or buttocks. Just hearing the thump of a beating in the principal's room would send shivers down the hallway. This would be a deterrent to any unruly behaviour, mischievous intent, and failure to complete homework on time.

Parents always sided with the schoolteacher, who was held in such high esteem that, when they found out of a beating, another awaited at home. Often, the father would administer it with his leather belt. Fortunately, I never received any beatings.

With clearly defined roles, the family structure was stable. Divorce and separation were unheard of, as were inter-racial and inter-communal marriages. These were confined within a community, with elders initiating matrimonial connections and cementing relationships. A marriage signified a bonding between families, and the arrangement was looked upon favourably.

A dispute between couples or adults was settled at friends' houses by elders in the community who enjoyed total submission and whose verdict and authority for any decision rendered were final. Such sessions were brief, and amounted to a lecture and dressing-down by the elders. This was followed by a meal as a sign that no one harboured a grudge or any animosity. Just to be castigated by an elder was sufficient punishment, since it put a black mark on the family's honour. Such was the interaction between adults and other communities which resulted in harmony and respect for each other, along with a little hidden fear that prompted people to tread cautiously.

The workplace atmosphere was clearly set along racial lines. The top leadership and managerial roles were restricted to the British and Europeans only, regardless of talent.

Indians were employed at clerical levels in the civil service and government agencies such as the railways and post office. They were all males, as females had not entered the workforce. They did not succeed beyond being a chief clerk or an assistant to a manager, but had the comfort of knowing that, as breadwinners, their income was secure to manage a household. Some who had ventured into business running small shops led an independent lifestyle. To run a small shop, only a simple permit—acquired over the counter at a post office or government department for two or three shillings (about twenty cents in today's- 2016 exchange)—was necessary. With such a permit, a shopkeeper could pretty well sell anything, since no regulatory inspections or enforcement systems existed.

Bicycle was the preferred transportation choice for commuting short distances, and most of us had one. They had a seat on the back to carry a passenger or tie a bag of groceries, and a bell to alert people in front. A headlight powered by a dynamo lit up when the bicycle was ridden. Cars were a luxury. We got our first car, a British-made Humber, in the late 1950s after we moved to Nairobi. Other makes available were Vauxhall, Austin, Buick, and Studebaker. Cars were standard shift with manually operated windows. Emergency gear included a spare tire that was either kept in the trunk or strapped on the roof. We also carried tire-patching kits to repair tubes when a sharp stone or nail punctured them, and chains to wrap around them in wet, muddy weather or when going uphill. There was no air conditioning, radio, or automatic options, and colour choices were either black or dark grey.

When tires were worn out to the rim, they would be retreaded. When they could no longer be retreaded, they would be sold to cobblers who would make sandals out of them for the Natives.

Turn signals were given manually by the driver: the left hand stretched out indicated a left turn, a circular motion indicated a right turn, and a ninety-degree bend at the elbow with the palm facing the front indicated a stop. When slowing down, the outstretched arm would be waved slowly in an up-and-down motion.

Turn signals were upgraded in later models with small lighted arms that would pop out by the left or right window to indicate a turn. Accidents occurred when a car rolled over or skidded into a ditch, but almost never with another vehicle or a pedestrian.

Regulated speeds were twenty to thirty miles per hour, and we rarely drove over fifty. We filled up with gas at *petrol stations* with a manual pump when in cities, or by transferring the contents of four-gallon containers into the tank using a funnel. Giving a ride to strangers was a common practice. It was considered extremely rude not to give a ride to anyone asking for it. We never feared a mischievous or criminal intent by a stranger amongst us. In fact, we would welcome such an additional person for safety, to help out in repairing a flat tire, or to push the car if it skidded into a ditch. Crime of any sort was very rare and never thought of.

The driving force of every parent was to encourage their children to acquire education as a key to a secure future. Swapping jobs to build a cross experience was not in practice, as longevity in one job amounted to loyalty and dedication to work—qualities that trumped all. Retirement took place at age forty-five; mortality hovered around fifty-five, and reaching sixty was a rarity.

The native Africans were relegated to menial office chores such as cleaning, gardening, cooking, and so on. Office orderlies would make tea for visitors, as that was a preferred choice of extending courtesies to them. In office buildings, hospitals, schools, railway stations, and even hotels, washrooms (called "water closets" or "WCs") were clearly labelled by ethnicity: Europeans, Indians, and Africans. The few buses that operated in Nairobi had the front seats reserved for Europeans, the middle ones for Indians, and the back ones for Africans. The train service that began in 1910 after the Mombasa-Kampala railway was constructed had similarly race-segregated cabins, washrooms, and facilities at their train stations.

Such was the system, with races living, working, travelling, and playing sports in their forced racial confines. There was no hatred, bitterness, or confrontation, since the system was ingrained, and there was no mechanism to redress a wrong.

Indians did not complain for fear of being labelled troublesome, which would lead to much worse consequences. But amongst the African population, there were slight, low-key murmurings about reclaiming stolen land. Apartheid was regulated, practiced, enforced, and followed to the letter, and everyone submitted to the law of the land.

In the late 1950s, Father transferred to the prisons services department, and was appointed the officer in charge of a Nairobi-area prison. This was a uniformed job carrying authority and prestige. He would not discuss his tiresome days at work, but we would hear him whisper to Mother how demanding his workload was, especially since the Mau Mau detainees overcrowded the prison's facilities. These detainees were undergoing an intensive re-education program from indoctrination of the Mau Mau oaths that had reduced them to sub-humans. The goal of many social workers, missionaries, and volunteers was to cleanse them of their oaths prior to releasing them into the general society.

Father Dwarka Dass, senior officer, Nairobi Prisons, 1957.

Father would walk five kilometres to work each way, a normal activity for many. On the few occasions when I had a day off at school, I would accompany him. Keeping a brisk pace in the early hours resembling a marching style impressed me. Seeing how much respect he got, with motorists stopping to offer a ride (which he never accepted), and in the camp as he returned salute after salute, made me very proud. My aim to follow a career in uniform was sealed.

A CAREER IN THE ARMY

Growing up in the bush where life was tough but simple, casual, and easy-going was in contrast to a city life, and it took me a while to adjust.

In the neighbourhood, a slight competition among the youngsters was evident, with people trying to outdo each other in material possessions. This would be with newer and matching clothes, as consumerism was slowly creeping in. Perhaps this was because of a higher disposable income in the well-paying city jobs? Fortunately for me, I always wanted to be in control of myself and do things as I judged them for appropriateness. I did not fall prey to following the majority and, as such, observed the excesses, but did not get drawn into them. For me, this was false, temporary, and meaningless competition. Neither did I ever felt left out or that I was missing anything.

Financially, I was not in the same league and I never looked at such a comparison as a disadvantage for not keeping up with my peers. A patch on a shirt, school blazer elbow, or shoe were inconsequential. Such visible displays of repairs did not cause me any uneasiness or lack of pride. They were minor issues that meant nothing to me and were never a concern to worry about.

Such trivialities were an unnecessary deflection from the real goals of life and, in later years, these very patches would be part of a designer clothes culture. I always followed a path of my choosing, stood my ground if ever remarked upon, and got comfortable setting a pace and leaving others to follow me if they so chose.

I was missing out on my rural, carefree living, and kept searching for an escape from the city culture. In high school, I was an active member of our Boy Scout movement. We would hold overnight weekend camps in school grounds and would spend the night in tents that closely reminded me of my earlier days. I would take part regularly in annual bob-a-job (a shilling for any job) fundraisers, which got me wearing my Scout uniform and knocking on business doors for any kind of employment. Most business owners were gracious, and would be pleased to part with a shilling or two for dusting off shelves, cleaning tables, or relocating files. They seemed interested in giving money, and took pleasure in watching a twelve-year-old work hard for a wage.

Such a vocation gave me the confidence to approach strangers in the knowledge that there was only a short window of a few seconds to present a worthy cause and plead for a job and not just a donation. Funds raised would be used by our Scout troop to buy basic training essentials such as compasses, blankets, water bottles, and backpacks.

In 1960 when I was sixteen, I attended an Outward Bound School course in Loitokitok on the foothills of Mount Kilimanjaro. This was an exhaustive three-week training session. It got us to climb the 19,342-foot peak, Kaiser Wilhelm Spitz, which was renamed Uhuru peak—*independence peak*—in 1962 to mark Tanganyika's—present-day Tanzania's—independence. The Outward Bound course with its motto, "to serve, to strive, and not to yield" and outdoor challenging environment provided a perfect introductory setting for anyone looking for a career in the military. I was to find later that such a course attendance was part of a cadet evaluation to identify leadership potential. The school would send detailed reports on each of us to our sponsors indicating our performance, motivation, team effort, initiative, creativity, and, most importantly, leadership qualities.

Our routine was that of a boot camp: rise at 0500 hrs, exercise in the frigid open for an hour, breakfast at 0700 hrs, room inspections at 0800 hrs, followed by classroom lectures in the morning and various presentations in the afternoon. In the evening after dinner, we would swap stories, tell jokes, and put on talent shows with singing and guitars.

Such a close environment where we depended on each other fostered camaraderie, bonding, and team spirit. The first two weeks were spent on acclimatization, building endurance, and learning field crafts and survival skills. We made our final climb in the third week, taking three days to climb up and two days to come down.

We started from the first base camp on the foothills at 0400 hrs, and reached the second base camp at noon and rested. The last leg of the climb, starting again very early at 0400 hrs, was bearable, as we had been well acclimatized and had built stamina to carry our backpacks of food, bivouac, and climbing gear.

The plan was to reach the edge of the crater by mid-morning, to walk around the crater's edge, and reach the peak by noon. At about 1300 hrs, we would start our descent.

I remember an incident walking on a narrow, loose gravel-covered slippery path along the crater's edge with a sheet of ice on one side and a drop of a thousand feet on the other, when, all of a sudden and without any warning, one of our group, James Johnson, started swinging his pick axe at us in a violent, wild rage. This was totally uncharacteristic of James, who had a charming personality and was a good entertainer at the camp fires. All of a sudden, we were his target. Fearing the worst should anyone slip off into the crater, we disarmed him, tied his hands, and held him to the ground until we all took turns to reach for the peak, which was a mere 100 metres or a thirty-minute careful walk away. We took photographs, and started our journey down with James resembling a chain-gang member in tow.

As we descended to the second base camp at 14,000 feet, James became angry with us and questioned why we had chained him, which had caused blisters on his wrists. He could not recall the events of the early morning and would not believe what he had done to endanger and become a threat to us all. Later, we were told that a lack of oxygen and altitude had affected him, a condition not unusual for some. But we were glad that no harm was done to anybody. James got over this episode reluctantly, and preferred us not to bring it up for the sake of his self-esteem and pride.

We respected his wish and he more than made up for it with his singing abilities at campfires and willingness to partake in team assignments. The adventure and experience of climbing a peak to 19,342 feet with basic gear was enriching for a sixteen-year-old.

The impeding decolonization process was slowly acknowledged by London, with Natives increasingly demanding the right to be masters in their country. The colonial government recognized that the wind of independence was sweeping across the Empire at a rapid pace.

When India became independent in 1947, other nationalistic movements across the British Empire with grass-root support were encouraged, and started pressing for political concessions. The Mau Mau movement slowly evolved from a regional terrorist organization to a nationalistic one demanding political representation. With such a momentum building up at a rapid pace, it was only a matter of time before Kenya would gain independence. Since my childhood growing up in remote Kenya, I had developed a deep love for the outdoors and an equally deep love for the African with his rich culture, traditions, and folklore.

Times were slowly changing, and talk about independence was commonplace and not feared for being labelled treasonable as before. In anticipation that independence would come in a matter of time, Britain started preparing for an eventual handover of various instruments of the government, civil service, security forces, and numerous agencies such as the railways, ports, airlines, meteorological services, and so on.

British expatriate personnel slowly started leaving, fearing a backlash from the Natives and possible nationalization of their properties. As such, opportunities for numerous jobs in the civil service, government agencies, and the army were opening up.

Uncertainty and panic ruled the day. Junior staff with extensive experience at lower levels were quickly promoted to fill in managerial and executive positions, with British personnel remaining alongside to guide, assist, and act as mentors. Unlike other European colonial powers and much to its credit, Britain had developed an efficient civil service and a functional and reliable judiciary.

Despite and against such volatility and uncertainty, my interest in enlisting in the army hardened, and I remained focussed on achieving my ambition. But I knew that it was going to be a tough call.

My aim to pursue a non-traditional career of an Indian in an African army in a colony of Britain started when I was still in high school. A city life was alien and the outdoor life was part of me from my early days in Ainabkoi.

First, my father had an influence on me. From his days as a nomadic forester moving all the time from campsite to campsite, small remote Kenyan town to small remote Kenyan town, made me love an outdoor life. Our close, regular interactions with the Natives brought me closer to appreciating their rich culture that had an established tribal structure embodying respect, and a well-defined hierarchical system.

I had won a scholarship to pursue a career in engineering in the United States. In the course of attending several interviews, I met the promoter of this initiative, a charismatic young politician, thirty-three-year-old Tom Mboya, from the Western Luo tribe, who had secured funding for Kenyans to attend universities in the United States. Mboya had been to the United States and developed close relationships with the church and many freedom-loving personalities who despised colonialism in any form. Mboya secured support for numerous scholarships for deserving students, and started a recruitment campaign to send students to universities in the United States.

I was to meet him again in later years. Securing a scholarship was reassuring, and brought into focus hopes of a bright professional future with the added comfort that, since my parents could not afford to pay for my fees after high school, I was not going to burden them. Father was the only breadwinner and we all knew that our finances were stretched since we managed our lives paycheque to paycheque. My interest remained in pursuing a career in the army. I never waivered.

A Kenyan or British soldier, with his colourful, impressive uniform, enjoyed much respect and lots of prestige. The army was looked upon as a unifying force across the country. It was also a key symbol of the country's pride since its marches, displays at annual agriculture shows in small towns, and regular recruitment drives complemented with corps of drums playing stirring popular marching tunes appealed to the public.

Opportunities were opening up to join the army and Africanization became a buzzword, with junior ranks rising fast; British officers started departing and they were gradually replaced by promising warrant officers (WOs) and non-commissioned officers (NCOs) who showed talent. Recruitment also began for direct-entry candidates straight from school. I was thrilled. It was going to be a long journey with much uncertainty, but I was full of confidence that, with hard work and determination, such lofty goals were achievable.

In 1962 when Kenya was still a colony, I applied for an officer cadet program whilst in my last year at high school at the Duke of Gloucester School in Nairobi. I did so just after writing examinations for a mandatory Cambridge University's Overseas School Leaving Certificate, which I passed easily and thus became eligible to attend an interview. An intense, multi-level screening process had to be passed for this rare opportunity, and competition was severe. The need for locally recruited officers for a new Kenyan army that would be formed after independence was huge, and there were simply no local officers to fill positions other than those who had been seconded from the British Army to the King's African Rifles.

A few African officers were the platoon commanders who were given the rank of *Effendi,* a rank created between an officer and a warrant officer. Some WOs and NCOs with experience leading soldiers were hastily promoted to officer rank, with British Army officers working alongside them and acting as mentors.

THE KING'S AFRICAN RIFLES: ENLISTING

I attended an interview in Nairobi a week after my high school graduation. I was one of thirty selected for enlistment in the King's African Rifles (KAR) from a field of perhaps 500. The evaluation course was for six weeks at the Military Training School in Lanet, in the Rift Valley province.

The KAR existed in East Africa from January 1, 1902, till about the mid 1960s, when they were disbanded, as the British Empire was winding down and colonies were gaining independence.

The regiment traces its origins to 1888 in Nyasaland, present-day Malawi, when the Arab slave traders ran a brisk business and frequently had confrontations with Natives who resisted the trade. About the same time, Britain was expanding its interests by acquiring new territories. The British Africa Lakes Corporation was an organization that developed economic interests and worked to eliminate trade in slavery. It recruited Natives, trained them like soldiers, gave them uniforms, and organized them into military-style units with ranks tasked to protect their remote stations.

They became the backbone of support to enforce order in the newly emerging colonial civil order as East and Central Africa were being colonized by Britain and Germany.

When Britain took control of Nyasaland in 1889, designating it a protectorate, the nucleus of the African Lakes security element and guards who, by now, had acquired sufficient training to be classified as soldiers, were designated the Central African Rifles (CAR), and given the unit designation 1st Battalion CAR. In a variety of enforcement engagements, the unit distinguished itself with its skills and was very successful.

A second battalion was raised, named 2nd Battalion CAR, and dispatched to Mauritius, the Gold Coast, present-day Ghana, and Gambia on the west coast of Africa. Their task was to provide security for the newly arriving colonialists and to enforce law and order. The units distinguished themselves with their loyalty, skills, and bravery.

In 1894, when Britain declared a protectorate over Uganda, Captain Lugard of the British Army, sent there to assist in providing security to the newly arriving British colonials, formed a military force, calling it the Uganda Rifles (UR). About 1900, the East African Rifles were formed in Kenya to support colonial civil powers in the newly emerging British East Africa protectorate.

The units were formed as battalions and were numerically numbered as they were raised: 1st and 2nd in Nyasaland; 3rd and 5th in Kenya; and 4th in Uganda. A 6th Battalion was later formed in Somaliland, the bulk of whose troops were from the former German East Africa.

On January 1, 1902, various battalions were amalgamated and declared the King's African Rifles.

The officer corps were provided by Britain with British Army officers, WOs, and NCOs seconded to the KAR. The soldiers of the KAR had an inborn flair for the bush, were natural trackers, were loyal, and, with their natural warrior instincts, easily submitted to regimentation. Their tribal background with many engagements with neighbouring warring tribes to settle grazing or cattle-rustling issues made them warlike and suitable for a military career.

The cadet training was rigorous, with a demanding schedule of physical fitness and classroom assignments.

We would get up at 0500 hrs daily, go on a route march of ten kilometres twice a week, return to camp for breakfast, and prepare for muster parade. Here, we would be inspected for cleanliness, haircuts, polished boots, and khaki shorts and shirts, which had to be heavily starched and pressed to finite creases. Mornings would be spent practising marching drills and weapons training. Afternoons were spent attending classes that covered subjects dealing with tactics and basic aspects of military regimentation.

Our instructors were from the KAR battalions from Uganda and Tanganyika and seconded British Army officers, WOs, and NCOs. Just before our arrival, one such instructor was Idi Amin, who rose to a sergeant in the late 1950s and was lucky to have been promoted an officer in the 4th Battalion, KAR, based in Jinja, Uganda. Amin was known as a troublesome, tough guy who was not easy to get along. He was not known to reason out an issue on its merits, but to be avoided as his way of enforcing discipline was by force. As an accomplished boxer, he looked at anyone and everyone as a potential foe in the ring.

We heard of many cases where he had resorted to violence to settle minor scores in camp, but reporting him was rare for fear of inviting his inevitable brutal fury. We would hear of Amin's cannibalism, but nobody talked about it.

I met him once on a visit to Uganda, as a guest of a former Sandhurst colleague, Gus Karugaba, at the officers' mess, Jinja Barracks, where I remember him calling me *karibu muhindi*—welcome, Indian! His lack of a basic courtesy to welcome a visitor warmly and thoughts of his cannibalism flashing through my mind made me nervous and, as dinner time approached, I took a hasty exit along with my host!

MONS OFFICER CADET SCHOOL, ALDERSHOT, UK

After completing Lanet, fifteen of us were selected to attend Mons Officer Cadet School in Aldershot, UK, for a six-month course in early 1962. Mons trained officer cadets for a short service commission primarily for National Service applicants who did not intend to pursue a career in the army. The academic and physical requirements of the course routine were demanding. The need was for Kenya to train and produce officers to fill command appointments at a rapid pace, and Mons's short course was the answer.

The instructors, who were WOs and sergeants of the Guards Regiment, Coldstream Guards, Grenadier Guards, Welsh Guards, Scots Guards, and Irish Guards, were not to be challenged; their authority was supreme and they had zero tolerance for the slightest of infractions, regardless of how minor or innocent. We quickly believed the widely heard saying that the backbone of the British Army was the WOs and sergeants' mess.

At MONS Officer Cadet School, Aldershot, UK, 1962.

A week prior to graduation, an offer came of three spots available in 1963 at the Royal Military Academy Sandhurst for Kenya, and those interested were asked to apply. A challenge here was to accept an imminent commission at Mons, with the graduation a week away, and start off with a career as an officer, or opt for yet another two years of demanding training and, in the course, lose out on two years of seniority.

I chose to apply, along with seven others, and this meant yet more rigorous tests and examinations before a place was confirmed.

Slight nervousness was inevitable for the remotest thought that Sandhurst, with its tradition of excellence, may prove beyond reach. An undergraduate university degree was preferable for eligibility, but not necessary.

Sandhurst had its own academic faculty staffed by senior experienced civilian professors who excelled in research, defence, international affairs, and war studies. Sandhurst did not award degrees, but graduates could work toward a postgraduate certificate in leadership and conflict studies. They could then complete a full master's degree via a distance-learning program from a university. Beyond meeting academic standards, a candidate had to pass an intensive selection process.

REGULAR COMMISSIONS BOARD, WESTBURY, UK

We were made aware that, to secure a place at Sandhurst, we had to attend and pass an intensive selection process at the army's officer-selection centre in Westbury, the only one of its kind in Britain.

Should we not pass, we would return to Mons and repeat the six-month training for a commission. Westbury had a reputation as a most demanding place and passing it was a prerequisite to enter Sandhurst. There were no exceptions. We had two months before attending Westbury, and were assigned to Wessex Brigade, based in Honiton, Devon, to keep us occupied and, I believe, for no other reason.

With determination that if others could make it, why couldn't I, I mustered courage and, still with some nervousness, was off to the Regular Commissions Board (RCB), Westbury, Wiltshire, an exclusive selection centre of the British Army for yet another week of intensive evaluation for suitability.

The RCB was based at Leighton House with forty acres of agricultural land around it. It had its own lake and a tough assault course designed specifically to test candidates.

The week-long selection process was thorough and fair. The range of tests and examinations had been developed over several years, and was specifically geared to select the right applicants for officer training. On the first day, we were briefed in detail on how the intensive selection process worked. We were told about the various tests, what was expected of us, and about the final assessment made on the last day, at the end of which the assessment board would determine our suitability as potential leaders.

The failure rate at Westbury was high and, hence, the drop-out rate at Sandhurst was zero since only the best-qualified and most promising candidates were sent there. On average, it took two attempts to pass an RCB. If we passed the evaluation at first go, we would secure a place at Sandhurst. If we failed on account of unsuitability or insufficient maturity, we would be recommended for remedial preparation prior to attending another evaluation after a year. And if we failed that, we would have to wait another three years. Such was the rigorous nature of the selection process.

The main evaluation process lasted four days. This consisted of physical, intellectual, and aptitude tests meant to assess applicants in conditions of stress and competiveness. Applicants were allocated a number by which they were identified and grouped in small teams of six to eight candidates. This group would be the basis of our team for the week, and we were expected to build bonds during this short period. We were to be tested on leadership under close supervision over a multitude of classroom and field assignments for physical and mental aptitudes. We were required to manage group discussions and lead teamwork assignments. Our individual personalities were evaluated to determine if we were suitable for officer cadet training at Sandhurst and as future military leaders.

Physical tests comprised of covering an obstacle course in a given time. For a group task, an applicant was given a leadership role on the spot that required allocating tasks to other members of the group and monitoring their efforts.

We were watched to see how we made a plan, how we interacted with other team members, and how we reacted to the weaker and stronger members. It was important to remain calm and focussed, and to encourage with compliments as a reward for input. Such a leadership role was under close scrutiny for a mere twenty to thirty minutes, but it had far-reaching consequences.

The classroom evaluation included a short essay chosen from a list of subjects, multiple-choice questions, and a five-minute presentation on a subject given on the spot for which no time was allowed for preparation. This was a premium institution that meticulously assessed candidates on a number of factors of suitability for leadership roles in the army.

There were trick skill-testing questions that had to be answered in group settings on the spot with no recourse or means to consult a reference. Our general behaviour, the way we walked, talked, and interacted at designated breakfast, lunch, and dinner tables with a different instructor sitting with us each time was under scrutiny.

There was no question that Westbury had fine-tuned its criteria to identify leadership qualities or that it had developed a well-respected selection process. It's no wonder that Britain's military leaders were meticulously selected with the right leadership potential for thorough preparation and grooming at Sandhurst for future senior military and policy-making roles to sustain the Empire.

THE ROYAL MILITARY ACADEMY, SANDHURST, UK

Much to my relief and joy, I was amongst the three who qualified for the spots that had been allocated to Kenya at Sandhurst for the intensive two-year course. One spot was allocated for the summer of 1962 and a colleague, Jackson Munyao, took it; the other two spots for a course starting in January 1963 were allocated to me and Bernard Killu, a fellow Kenyan from the warrior Kamba tribe.

We lost out on the two-year seniority that would have come with a commission at Mons, but were glad to have had this opportunity and the anticipation that, in time, it would reward us with higher command appointments. Hence, we started our training at the Royal Military Academy Sandhurst, UK, in January 1963.

Attending the Royal Military Academy Sandhurst, commonly referred to simply as Sandhurst, has been a privilege. Sandhurst is Britain's premier officer-training centre for the British Army, founded in 1947 through the amalgamation of the Royal Military Academy Woolwich, which was founded in 1720, and the Royal Military College Sandhurst, ranking among the tops globally. It was a boys-only college until 1984, when women were admitted.

Training at Sandhurst developed an individual's character and built on teamwork, problem-solving, decision-making, and negotiating skills. This was a key preparation for the future officer to communicate effectively and confidently with his soldiers and get the best out of them.

Over several exercises, a cadet is exposed to complex situations requiring a thorough analysis that weighs pros and cons, and the ability to make decisions with flexibility as the situation changes. Emphasis on leadership is evident throughout.

The academic part of training was supported by the departments of military and war studies. This included an understanding of campaigns, theories of war, internal security, and counter-insurgency operations in the context of political and strategic interests. The department of international affairs focussed on the role of the armed forces in a political and strategic context.

Throughout, all courses gave an insight into human behaviour, motivation, team-building, communication, problem-solving, creative thinking, and leadership, with an emphasis on shaping the thoughts and actions of an officer in a leadership position.

The two-year program was split into six terms of about fourteen weeks each; cadets advanced from junior to intermediate to senior divisions and they were identified by different coloured lanyards. The initial six-week basic course was much like a rigorous boot camp that vigorously breaks down a cadet's carefree civilian personality. This was then moulded with intensive drills and field exercises so disciplined and regimented they often made weak minds wonder if they'd made the right career choice.

Cadets were selected and appointed to manage other cadets via a ranking system that saw them get ranks of corporal, sergeant, junior under-officer, and senior under-officer. Cadets with such ranks exercised authority to the fullest and handled minor infractions with zeal and by the book with no allowance for leniency. They took every advantage to enforce their leadership, knowing too well that they were being tested, and so flexed their muscles to the fullest. Cadets were interviewed for their regimental choice in the intermediate term of the second year and competition by the cadets to join a regiment of their choice and be selected by the regiment as the most qualified was severe.

Sandhurst had its own brass band which supported barrack drills and played at the graduation parade, the Sovereign's Parade. This was held outside the Old College on the parade grounds with invited dignitaries, friends, and families of the graduating cadets in attendance.

The reviewing officer at a Sovereign's Parade would either be a member of royalty or a senior military officer representing the sovereign. Sandhurst represents the best of British pomp and circumstance for tradition and ceremonies. The Sovereign's Parade would mark the graduation of officer cadets into commissioned officers, and the event is the best of the best in British military ceremonial excellence.

The event would run as a textbook example with clockwork-like military precision, everything happening to the minute, rain or shine.

The reviewing officer would inspect the parade company by company, with the brass band playing stirring military tunes. The parade would include the trooping of the colour, which would march past the reviewing officer with the graduating senior division leaving the parade grounds via the grand entrance of the Old College, followed by the academy adjutant on a white horseback.

Awards for the best cadets judged by the commandant would be a sword of honour for the best cadet joining the British Army, a similar sword for the best overseas cadet, and a Queen's Medal for achieving excellence in academic work. Guests of the graduating division would be invited to a formal lunch in the cadets' dining room. An impressive commissioning ball would follow and, at midnight, the newly commissioned second lieutenants would proudly display their "pip," a rank insignia that, for the first time, bestowed the privileges of being saluted by junior personnel and addressed as "Sir."

The commandant at Sandhurst was a major general. The commandant on my arrival was Major General George Gordon-Lennox, who was succeeded a year later by Major General John Mogg.

My two years at Sandhurst were full of adventure and unique experiences that no other institution would have offered. I had come from a colony where the recent events of Mau Mau were still fresh and daily news coverage of rapid advances toward independence in the colonies was overwhelming. In time, it would dawn on me that being of Indian ancestry and coming from an African colony would reflect on the makings of a new multi-racial Kenya.

I was enrolled in Blenheim Company, Old College. I was allocated a room in the dormitory that I shared with a fellow cadet, Andrew Keelan, and assigned an orderly, Sid, a cheerful, friendly character of Irish lineage who had been there for many years and was accustomed to seeing cadets come and go. He would be responsible for keeping our room spotless and would take care of our laundry. This chore relieved us of an important responsibility such that we could concentrate on using precious time thus saved on studying.

Officer Cadet, Blenheim Company, Old College, Sandhurst, 1963.

It also reflected on a kind of exclusiveness for the future officer to get accustomed to having his routine chores taken care of by others. After a welcoming week of orientation, it was clear that there was not going to be any room for skipping classes, submitting assignments past their due date, being late for a class or ungentlemanly conduct whilst on or outside Sandhurst grounds.

On Sandhurst grounds when in civvy clothes, we had to wear a hat and tip it every time we passed a lady. Our three-button, navy blue, double-breasted blazer had to be kept buttoned. Wearing sports clothes was only allowed when going directly to the tennis courts or gym, or when playing a team sport such as rugby, field hockey, or soccer.

Crossing the parade grounds for whatever reason was a major sin and to be avoided, with extra guard duties awaiting as punishment. The parade grounds were sacred and only to be used for parades.

We could commute in the vast grounds on the army-issue, olive green-painted bicycles available to rent. Riding a bike, saluting an officer while on it, and parking it had specific drills that had to be followed to the letter. Sandhurst had its own rifle range within its grounds where we practised shooting with the standard-issue 7.62 mm rifle, 9 mm Sten gun, and 9 mm pistol on a regular basis. There was also a rigorously designed assault course that we had to go on once a term. Lots of sports facilities dominated the grounds.

There were tennis courts, field hockey and cricket arenas, soccer fields, an Olympic-size swimming pool, horseback riding, and canoeing; indoor sports included squash, fencing, and boxing. Outside subsidized club membership was available for those interested in golf, sailing, flying, parachute jumping, or any other sport.

The environment was rigorously academic. The library was well stocked with a rich reservoir of reference books that introduced military tactics and discussed and evaluated previous campaigns' achievements and failures.

The most important part of such research was to study the commanders who planned campaigns, what their strategies were and how they were influenced, how they performed and instilled morale in the troops, and how tactics evolved over time with new weaponry. Whether tactics should evolve around new weaponry or if weapons should be developed to suit a doctrine were subjects of ongoing discussion that was beyond some of us from the colonies whose knowledge was limited. It was clear at the outset that, to catch up with the other 170-plus cadets, some of us had to work really hard to benefit from classroom discussions by contributing and absorbing or we would simply become irrelevant. What we did not want to do was be labelled cadets who were there on political or diplomatic considerations. We were there to participate fully to the best of our ability with determination and confidence, and to be judged along with others on an equal footing.

FIRESIDE CHAT: JALLIANWALA BAGH

The academic part extended from classroom to semi social. We would be invited in small groups for fireside early-evening discussions by the director of military history, the well-renowned historian, the late Dr. John Keegan, at his house on the Academy grounds. Our company instructor, Captain Geoffrey Inkin, attended some presentations, obviously with an aim to get insight into us. It was up to us to select a subject, seek approval in advance, and make a five-minute presentation followed by a discussion. We were quite aware that, aside from the social overtones of such a presentation, we would be judged on how well prepared we were and how we handled the discussion and drew conclusions. This was yet another reminder that no matter what we did, we were closely monitored.

On one such evening, I chose to speak about the *Jallianwala Bagh* massacre in India. This took place in 1919, on the outskirts of Amritsar, not far from my father's ancestral village of Hoshiarpur in the Punjab. The incident was close to home and was recalled with vivid sequence by many elders and some who had witnessed the shooting from outside the wall.

Over time, the younger generations would hear of these atrocities and they became a major topic of discussion, fuelling further animosity against the British since many family and close friends became the innocent victims.

The subject was reflective of a volatile political environment but also of military interest since the army had been sent in to quell a possible disturbance if the gathering got out of hand and became violent.

I selected this subject as we would hear horrific stories from Father at the dinner table about the brutality that Indians endured during British Raj.

In retrospect, it would have been wiser to have selected another topic that was less confrontational. Over 20,000 innocent, unarmed nonviolent protesters, some of whom were illiterate farmers who had arrived from neighbouring villages, gathered in a public gated park, the *Jallianwala Bagh*. Their aim was to celebrate a festival and protest against the arrest of some prominent freedom fighters.

The park was around six acres, and measured about 200 metres square, gated with high walls and five sealed entrances. There was a well in the centre. Apparently, a curfew was in place, but it was not widely advertised and the villagers from outside were unaware of it. After warnings, the crowd failed to disperse.

The British commander, Brigadier Reginald Dyer, ordered his fifty or so Gurkha troops to open fire against the gathering. Since exit gates had been sealed, many were trampled or shot, and some jumped into the well rather than be shot or captured by the British Indian troops. General Dyer's orders were to shoot and kill the protesters and any escapees. The firing lasted nearly ten long minutes and only stopped when troops ran out of ammunition. With varying accounts, over 1,000 were killed and another 1,500 wounded. In addition, General Dyer sealed off the street leading to the main entrance to the park. Since residents on the street had no back doors, they were forced to use the road by crawling on their knees, face down.

With a curfew in place, they could only go in and come out during the day. The brutality bordering on barbarity stunned the nation, and drew widespread anger.

This incident further deepened hatred against the British and by some historians' assessments, speeded the nationalist momentum for India's independence. Surprisingly, Dyer, rather than face a court martial for his reckless conduct, was praised by the ultra-conservative British House of Lords for his actions but renounced by the British House of Commons with Winston Churchill, secretary of state for war, calling it a monstrosity and for the immediate resignation of Dyer.

To the disbelief of many, right-wing conservatives raised over a million pounds (in today's figures) to help Dyer. Flashes of the Hola massacre in Kenya in the aftermath of the Mau Mau rebellion would also come to mind.

The *Jallianwala Bagh* incident marks a prominent milestone in India's struggle to gain independence from the British, which it got in 1947. Such an event, amongst many others, showcased Britain's cruelty toward its colonial subjects, where brutal force was used to enforce submission. Such events also reflected Britain's divide-and-rule policies which pitted neighbour against neighbour. Caste and religious differences were played to the fullest to keep people divided and win favours. Administration of justice and unchecked law-enforcement excesses were often left to the discretion of senior British officers on the ground. They exercised such unchecked authority with utmost zeal, knowing that they would not be held accountable for their actions.

During my presentation, I used the term "freedom fighter" a couple of times with reference to the non-violent protesters. I maintained that the protesters were not armed, did not cause any property damage, were not unruly, and posed no threat. They were clearly demonstrating their right to air grievances against perceived injustices. They wanted to be masters in their own home. But now they became hostages in a confined, sealed yard with no chance of an escape.

I argued that General Dyer's conduct was most unbecoming of an officer, as he had several options that didn't include the use of force.

He could simply have kept the yard sealed, identified the organizers of the gathering, and arrested them if a permit was required for such a gathering.

But the gathering was also a celebration of a spring festival, the *Vaisakhi,* and such impromptu assemblies without having to take out a permit were part of an annual celebratory culture, practised over the previous several years. The incident brought out the fundamentals of officership when deployed in aid of a civil power mission. Leadership by simply giving orders to fire upon unarmed innocent civilians is certainly not the way to win the hearts and minds of the civilian population. John Keegan complimented me for my presentation, which discussed the clear military and political aspects of the event, but politely reminded me that a cadet should not be heard to discuss political matters as such views were not to be harboured by military officers. Wonderful advice. My relationship with John Keegan remained warm and friendly despite my choice of a confrontational topic with military and political overtones. I would often consult him on many a course assignment for guidance. He was always gracious and willing to assist me.

I should have remembered that during my earlier times with the KAR, we were forbidden to discuss or hold opposing political thoughts. I was to hear later that John Keegan, by now Sir John Keegan, went to West Point Military Academy in the United States for a year on an exchange program and left his house closed.

On his return, much to his dismay, winter freeze had taken its toll: pipes had frozen and burst, and the water damage was extensive. The house remained unused as it was too expensive to renovate, and John lost a precious collection of heirlooms and memorabilia.

As for physical aptitude, we were expected to ride a horse as the British officer class enjoyed a privileged social standing. They would enjoy membership to exclusive private clubs where bowling, cricket, tennis, and horseback riding were the norm. Very few select outsiders had access to such privileges, which were restricted only to those who were guests of a member.

This was a life lived in a bubble. We were familiar with the private clubs in Kenya that we never visited or sought membership in since, with a racial divide-and-rule policy, we did not belong there and were not allowed in. I was to discover that a private boarding school tie and the right accent carried a status label which opened doors to opportunities. The deeply entrenched British class system where lineage automatically identified social standing came to the fore. I had never ridden a horse and knew anything about them.

HORSEBACK RIDING

One of the first activities we had to learn was to ride a horse. Early one morning, I was assigned "Jack," and given brief instructions by the stable sergeant, Sergeant O'Conner, on how thighs, legs, feet, and stirrups would send commands for the horse to obey.

After kitting myself with proper gear, I nervously mounted Jack. He slowly walked along a well-trodden, narrow path in the woods. Fearing that my knees were going to hit a tree, I panicked and let go. I fell, and Jack took off to return to the stables. I limped there, too, and was greeted by an irate sergeant. He barked orders to the effect that I did not follow instructions or take care of the horse, and that I was negligent in letting him go.

He was not concerned about my state and didn't ask if I was hurt or needed medical help. I was let go and told to appear before the company commander, Major John Andrews, at 0800 hrs, the following day.

I was marched in and the charge sheet was read out loud with three serious violations: not taking care of government assets entrusted to my care, not exercising a leadership role on the horse, and bringing disrepute to a privileged sport. When asked if I had anything to say, I kept quiet and firmly replied, "No, Sir." But inside me, I was controlling myself from bursting out with amusement over this accusation, regardless of how well intentioned it was to instil a measure of accountability.

I got a weekend suspension of privileges, which meant that I was confined to barracks for a weekend. This was a blessing in disguise, as I was able to complete my classroom assignment ahead of everyone and, happily, was graded "above average." Captain Inkin sympathized with me, saying that I was henceforth taken off riding sessions and instead could focus on other sports.

Sandhurst exposed us to the world of banking. We had our bimonthly pay of fifteen pounds deposited in our local Sandhurst bank. We operated our account by writing cheques and had to make sure that there were funds in the account to cover them. An NSF cheque meant the individual was incapable of managing his finances, and therefore was unsuitable for a commission. These cheques were valid only at Sandhurst facilities such as the gift shop, barber, tailor, cobbler, and bar. Outside, most of us had an account, often a savings account at the post office, which was standard practice in those days.

Sandhurst provided numerous opportunities to build lifelong relationships. With its role in preparing leaders, it had rightly acquired the status of an exclusive training facility. It was an obvious place of choice for sons of leaders from Britain and many countries to attend a two-year, intensive cadet-training program leading to a commission in the army. Entry was just for males, as females served only in the select few women's auxiliary units and were trained elsewhere at women-only centres. As all careers opened up to women, they started attending Sandhurst in 1984. We had two intakes a year of around 170-180 each, split into three colleges—Old, Victory, and New—which were further split into four smaller units per college called companies, each with around fifteen cadets.

This small number of cadets and the high ratio of staff meant a cadet had about two teaching, training, and mentoring staff who closely supervised, coached, and instructed him. Our two-year association with fellow cadets with whom we lived, studied, played, and worked in even smaller groups on projects and assignments requiring collective input, and got us to develop close bonds.

These bonds enriched us with a cadet's lineage; some of our ranks were descendants of important figures whose contributions were of a historical significance. Such contacts provided a bridge into the past.

FRIENDSHIPS AT SANDHURST

One such cadet was Jonathan Orde Wingate, who came from a distinguished family and whose company I enormously enjoyed. We would go to pubs and share stories, and developed the kind of closeness that naturally comes with common interests and time spent together.

Jonathan would share his background with us and, at times, we would think that he was lonely and somewhat detached, despite carrying a well-known name in military circles. Jonathan was the son of the famous Orde Charles Wingate, who earned a place in history with his relentless bravery, creativity, and commitment to a cause.

He formed the well-renowned Gideon Force in Sudan in the 1930s to fight the Italians and lead long-range penetration units called *Chindits* in Burma during WW II to fight the Japanese. He played an important role with the Jewish diaspora who were facing persecution in many countries and had started planning to create a country of their own, eventually to be called Israel.

His grandfather was the governor general of Sudan, his uncle was the governor general of Malta, and his other uncle was a judge in the British courts. Despite this rich ancestry, Jonathan remained somewhat aloof to many, but became a close friend of mine and, happily, took me into his small inner circle.

Jonathan was an only child and sadly, never met his father, who was killed in an air crash in Burma just six weeks before he was born in 1944. He did not flaunt his famous name, maintaining that he had to earn his place with his ability and competency. He would tell stories of what he'd heard from his aunts and uncles at dinner time, stories we were to read later in books. His father was an eccentric who followed establishment rules with a rebellious mind. He challenged his superiors forcefully and without fear if he felt that his suggestions for a plan were better.

Such a rebellious streak coupled with personal hygiene issues, as he was often scruffily dressed, meant he stood out in a group. He would carry garlic around his neck and was known to bite on raw onion for health reasons, regardless of whether he was in company. He was seen as troublesome but was respected just the same for his bravery, leadership, and creativity for the new tactics he introduced to improve the army's doctrine.

But his ingenuity, selfless attitude for a simple lifestyle, devotion to his cause, and love of men earned him credibility right up to the ranks of Winston Churchill and Field Marshall William Slim of the 14th Army. Jonathan was most proud that his father had trained the famous one-eyed Moshe Dayan with whom he formed the successful guerilla force, Special Night Squads, the forerunner to the Haganah. They carried out successful small assaults to fight the Arab revolt as the State of Israel was being set up. It is widely believed that Orde Wingate's strategy to lead from the front, and to take the war to the enemy with long-range penetration patrols at night, covert operations, and ambushes, formed the doctrine of Haganah and the Israel Defence Force.

Jonathan would recall with fond memories that Israel's first president, Chaim Weizmann, was his godfather, along with Emperor Haille Sellasie of Ethiopia.

He would also reflect on his distant relationship with T.E. Lawrence, the famous Lawrence of Arabia. Jonathan joined the artillery and we communicated with each other till about the mid 1970s. On one or two occasions when I met him in London just before he retired, he appeared exhausted and somewhat aimless.

The demanding rigours of an army life did not interest him any more. I noticed that he took to complaining about minor incidents endlessly, and perhaps was trying to justify that a more rewarding life awaited him in civvy street. After an early retirement as a major in 1978, he went into private practice as a head hunter and I did not follow further on how he made out.

I lost contact with him, but would hear that, in later years, he became a joint patron of the Chindit Old Comrades Association with Prince Charles.

Sadly, Jonathan passed away in 2000, and *The Daily Telegraph* of November 3, 2000, carried an emotional obituary.

Another cadet whom I befriended in my intake was Panji Kaunda, son of Kenneth Kaunda, the president of Zambia who ran the country for over a decade. Panji was not destined for a career in the military, and it was obvious from his attitude and general behaviour that he was sent to Sandhurst under duress. We did not keep in touch, and I never heard how he made out on return. Prince Naif Ali of Jordan was also in our intake, but we did not keep in touch.

I have also been privileged to have maintained close contacts with some inspiring senior officers whom I looked upon as heroes with utmost pride and who, in many ways, influenced my future. In particular, the commander of the 70th Infantry Brigade, the King's African Rifles, was Miles Fitzalan Howard, whom I first met when undergoing basic training in Kenya and who would visit us at Sandhurst. Such infrequent and short visits would lead to a warmer personal relationship in later years. He inherited the dukedom of Norfolk, operating from Arundel Castle in Sussex. I exchanged numerous letters with him and paid calls on him both at the castle and at his pied à terre in Euston Square, central London. His unconditional offer to me to use his name as a reference should a need ever arise was most humbling. It was unthinkable for anyone in Britain, seeing his name as a reference on an application for employment or any other opportunity, to reject it.

In later years when I discussed my future plans to migrate, he encouraged me to consider Britain as a first choice, and volunteered to assist me in getting a job with a government agency. It took me a great deal of courage and effort to politely decline such a generous offer, as Canada was becoming more attractive.

Another officer I befriended was Brigadier John Hardy, who commanded the new Kenya Army at its formation in 1964. I kept in touch by visiting him at his residence in Wimbledon whenever travels took me to London. He was always gracious with a warm welcome and generous hospitality extended at home over many a delicious curry lunch.

He was a fine gentleman who had a warm, friendly personality and was always there to encourage and guide me as I was advancing in my career and, later, when I had migrated to Canada.

Libya

Whilst at Sandhurst, we travelled to several places as part of our training. Such travels widened our horizons and exposed us to different climatic regions, cultures, and people. It broadened our outlook, giving us the confidence that a military operation was within reach in far-flung regions. We were fortunate to learn such techniques early on since the British Army, whilst sustaining the Empire, had a global reach and, with its many obligations to treaties and membership in alliances such as NATO, it had to have units well trained to go into action in any part of the world at a moment's notice.

Such training introduced us to how various parts of the army such as tanks, engineers, artillery, signals, and medical units worked. From classroom academic exercises and Tactical Exercises Without Troops (TEWTs) conducted on sand models, participation in actual training theatres using live ammunition was educational and most rewarding.

Happiest time for a soldier – letter from home.
Receiving it on an exercise in Wales, 1964.

Such training extended to inter-service operations, setting up and running command-and-control centres, and learning how operations with the air force and navy were coordinated.

In March 1963, we went to Libya for a week, staying first at the American Wheelus Air Base, outside Tripoli. The base was a replica of a town in the United States. Its street names and base layout, and string of stores, fast-food outlets, ice cream parlours, cinemas, and restaurants, and where the currency of transaction was the greenback. It was difficult to believe that one was not in Arizona or Utah.

We participated in an exercise with a British Armoured Tank unit, deep in the Great Libyan Sand Desert just south of Benghazi, in the Cyrenaica region.

Britain maintained numerous training facilities around the world in varying climatic conditions where military units were trained regularly in readiness for deployment at a moment's notice to far-flung places.

It had training facilities in Libya, where units would undergo desert warfare training on a rotational basis. The desert experience was very demanding on endurance. Soaring daytime temperatures and bitterly cold nights gave us an insight into how tough life is in the desert, with its numerous challenges for survival in everyday conditions, to say nothing of fighting a war. Availability of water was perhaps the biggest need all the time. A few scattered settlements of nomadic Bedouin travellers were by an oasis or a dry riverbed, a *wadi*. Here, water could be reached by digging a small, shallow hole with a shovel.

On one such exercise, we had to cover about fifteen kilometres on foot. We started from base around 0400 hrs. By 1000 hrs, we were resting in tarpaulin-covered shelters till late afternoon when the sun set. Doing anything physical during the day was just impossible, as the heat and scorching sun were unbearable.

I used to hear that one could fry an egg by simply breaking it and putting it on the spare tire dip on the hood of a Land Rover. Sure enough, I tried it and, after about fifteen minutes, the egg was indeed bubbling. One just has to imagine the amount of heat required for such a task and the extreme hardships that the nomadic camel herders in their quest to trade and survive endure as a lifestyle on a daily basis.

HITCHHIKE TO EL ALAMEIN, EGYPT

Summer breaks at Sandhurst provided us with an unparalleled opportunity to pursue our instincts to the fullest.

An eight-week break was just too much to sit around and do nothing or pretend that one was going to study, as that was not going to happen. Bill Williams from Victory College and I decided that we should embark on something unique and extraordinary: visit the battlefield of El Alamein in Egypt where Montgomery ("Monty") with his Desert Rats had beaten Rommel's Afrika Korps during World War II.

Bill had a pleasant personality and, with his Irish lineage, was a natural entertainer and the spark at a gathering. He also viewed regimentation as an unnecessary inconvenience and although he submitted to the tightly structured routine, he would always find ways to circumvent it, if he could. We would visit pubs on Saturdays and I found his company stimulating. He would be a pleasant, jovial partner to travel with, and I thought we would make a great team.

The Battle of El Alamein was a project on which I had worked as part of a term assignment. On one of the lecture series, Monty talked about leadership, and listening to him was inspiring. A short man with a soft voice, he was completely devoid of the gung-ho combat commander's personality depicted by Hollywood, and was breathtaking.

Monty was a battle-hardened hero who was widely respected for his accomplishments. Just listening to him in the same room was extraordinary and inspiring. Within his outwardly warm personality, he was tough, with nerves of steel. Once he'd made a decision, he would defend it to the limit without fearing backlash from his peers or superiors. The substance of his speech, and how he articulated simple fundamentals of high command, from maintaining aims to meeting political goals, was inspiring and motivating. His focus on assessing enemy strengths and weaknesses and on planning operations in a sort of textbook-type order that gave priority to that most important element, the soldier, was overwhelming.

He not only selected his field commanders with utmost care, but emphasized that every soldier right down the ranks know the bigger picture in which he was operating with his superiors' full confidence of his ability. He was known to relieve commanders at will who did not measure up to his exacting standards. He could also be confrontational to those peers and superiors who did not give proper consideration to his opinions or suggestions to improve a plan.

He also had a reputation for being rude, arrogant, and uncompromising. In our young minds, these were the qualities admired in a senior commander who stood his ground and remained focussed on achieving his mission without fear of repercussions from superiors. The talk captivated the audience, and I became interested in visiting the scene of battle where he commanded the mighty Eighth Army for which he had earned wide publicity. Hence, this is why I chose to visit El Alamein.

The Axis forces of Germany and Italy under Field Marshal Erwin Rommel had secured a base in Libya, and were advancing east with the aim of capturing the Suez Canal to disrupt the Allies' oil and other logistical supports. Lieutenant General, later Field Marshal, Montgomery, assumed command of the Allied Forces under the 8th Army consisting of troops from Britain and the Commonwealth.

"Monty," as he was widely known, acquired a distinct cult-like personality, with soldiers eagerly following him whenever and to wherever. He wore two cap badges on his beret, one of a tank regiment and the other of the British general officer, a unique distinction that he carried with pride. His tough personality and frequent pep talks to soldiers in the frontlines strengthened courage and instilled morale. The Allied powers' goal was to advance west from Alexandria to Tobruk in Libya, and Monty chose to reinforce a defence line at the coastal town of El Alamein. This defence line extended around 100 kilometres south to the inhospitable, desolate Qattara Depression, which lies some 100 metres below sea level.

We had very little money from our meagre salary. But with our vision, determination, and high ambitions, we decided that we would embark on our trip by hitchhiking—yes, hitchhiking. Many thought we were out of our minds. We made an outline of a plan: we would go by rail to Dover, cross the English Channel by boat to Rotterdam, Netherlands, and offer our services to the Shell Oil company. Our request was to go on one of their oil tankers to Egypt and back in return for general manual labour on the ship. We had researched that ships required labour for all kinds of unskilled work.

We approached the academy adjutant with our plan by presenting a scholarly objective to visit the battlefield of El Alamein to study Montgomery's successes, maintaining that our mission had relevance to our classroom project.

We were not sure if we needed permission, but felt comfortable to have advised the academy since we were contacting a foreign diplomatic mission and a major commercial enterprise. As a further safeguard, we felt our movements should be known to the academy just in case we ran into trouble and needed help.

An endorsement from the academy would also open doors with the confidence that our trip was officially sanctioned as a worthy project. Happily, we got permission to go ahead, and were offered assistance that facilitated our ambitious adventure.

The academy made contact with Shell's head office in London who put us in touch with their Rotterdam office. The academy also contacted the Egyptian embassy in London. Our request received almost an immediate acceptance, and we were offered two places on several sailings to and back which, on average, were once a week. This news gave us a great boost.

Brigadier Shazly, the military attaché in the Egyptian embassy in London, was most helpful in offering assistance to us once we got to Port Said. We had mild concerns that, with Egypt having recently taken over the running of the Suez Canal in 1956 from Britain under acrimonious circumstances, we may encounter difficulties. But this did not happen.

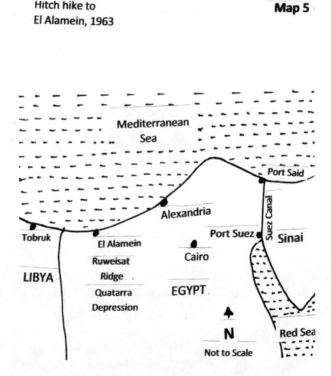

Hitch hike to battlefield of El Alamein, Egypt, 1963.

We started our adventure on July 15, 1963, and got to Rotterdam in the mid-morning. Our backpacks contained a bivouac, sleeping bag, several hexamine tablets for an army-issued field stove, and a basic change of clothing that included a necktie. After formalities, we were assigned to an oil tanker that was leaving port that afternoon.

Our five-day sailing took us around Gibraltar and the Mediterranean, arriving at Port Said on July 20. Our contact was the executive officer, Captain Fitzgerald. An ex-Royal Navy officer, Fitzgerald was a warm, friendly person who briefed us on the ship's routine and our workload therein. We were to be the deckhands, charged with scrubbing and cleaning the surroundings of the small swimming pool. We had to ensure that the deck furniture was taken out of storage every mid-morning and placed by the pool and removed by sunset and stored. At 1800 hrs, we would assemble in the Ward Room for a drink and sit down for a multi-course dinner, formally served. We were treated like officers, and had to wear a necktie to dinner. The other officers were just as welcoming and friendly, but curious as to why we would opt for such an arduous adventure. The five days passed by very quickly.

We arrived in Port Said mid-morning and, much to our surprise, a hat had been passed around and we were given a handsome tip for our work. Twenty pounds between the two of us was almost a week's wage. We got a ride in a truck to Cairo courtesy of the port supervisor and were dropped off in a bustling, lively downtown. We had no idea where we were going to spend the night, and hotels were too expensive and out of reach. We had no contacts other than the address of the British Embassy in case of an emergency.

With evening approaching fast, we thought of a novel idea. We decided to go to the police station to ask if we might be allowed to pitch our bivouac in their compound. We were strangers to the city and on meagre financial resources. As cadets, we thought we may enjoy a preferential treatment. The officer in charge, Superintendent Abdul, perhaps aware that he was not allowed to and with night approaching fast, took pity on us. After checking our credentials, he showed us where we could pitch, and even allowed us use of bathroom facilities. Overjoyed with such a warm offer, we had a dinner of soggy sandwiches and dry biscuits which had been given to us before disembarking the ship. The weather was unpleasantly hot and humid. We did not dare make a fire to cook in a police compound, and so just had to rough it out.

Early the following morning well before sunrise as his shift was changing and he was signing off, Superintendent Abdul offered to take us home.

After breakfast, he dropped us on the Cairo-to-Alexandria Road where we could continue hitchhiking. It was a wonderful offer that was hard to refuse. We were warmly welcomed to their beautiful house. After a delicious breakfast, his wife, with motherly instincts, packed generous portions of meals and Superintendent Abdul took us to the main Cairo-to-Alexandria road. Within minutes, we were on board a cattle truck, standing at the back in the utilities enclosure just behind the cabin with perhaps five cattle and some goats. Luckily, the driver was going past El Alamein, a distance of some 250 kilometres from Cairo. A four-hour drive in midday hot sun and a cool coastal breeze got us to El Alamein, and we were dropped off by the roadside in the mid-afternoon.

With scorching heat and in the open, we took refuge in a bus shelter until the day cooled. Evening was setting in and our anxiety built regarding where we would spend the night. By about 1900 hrs, we walked some 200 metres inland over rocky, sandy ground, pitched our bivouacs, and had whatever dry rations we were carrying. The night was cool but pleasant. We were up at 0600 hrs, woken by the slowly increasing heat. We walked back to the coast, as going further inland was asking for trouble. The desert heat was going to become unbearable, and a cry for help or rescue was out of the question.

Our goal was to visit Ruweisat Ridge, fifteen kilometres to the south, which was a key reference feature of the Battle of El Alamein. And it was here where a young Lieutenant Charles Upham of New Zealand had won his second Victoria Cross, the only one to be awarded this most prestigious award for gallantry during World War II twice. New Zealand later honoured him by naming streets, a ship, and an aircraft after him. We could see the hazy outline of the Ridge. We tried our luck with the police again. A good, kind-hearted policeman came to our rescue and, in a short while, we were with a small caravan of camels who were heading toward Ruweisat Ridge and beyond. Riding a camel was an experience. It was very comfortable sitting atop that animal in a basket and going at a slow, almost ceremonial, swaying pace.

The Ridge was like any ridge, with hardly a remnant of war in sight. We were in absolute desert wilderness. We got off at a small caravan campsite where travellers would spend the night. At the foothills of the ridge, there was a small, shallow pond not unlike an oasis: damp, but with almost no water. Two camel drivers dug a small hole with their shovels and slowly collected water and made tea. It was a refreshing experience that proved that nature has a way to avail extraordinary means of survival. Being with total strangers and suffering language barriers was somewhat troubling. We were at their mercy. It was fair to assume that since the caravan operators had so willingly agreed to assist us, they were known to the police. As such, they would not inflict any harm to us.

Early the following morning, the caravan continued on its way and, by hand signs, we understood that another caravan would stop by in a day or two with whom we could return to El Alamein. The sheer thought of being alone in the desert with no war relics to visit was a little daunting and made us wonder if we'd end up being victims of an attack. Such thoughts occupied us the whole night and every little noise of any sort was unnerving. We were alone the following day and the midday heat and scorching sun were unbearable. The only shade was in a shallow, cave-like opening, but we had heard about snakes and scorpions—we remained on guard and quite nervous.

The second night on our own was intimidating. We questioned what had made us embark on such a difficult trip, and were convinced that even foolishness and stupidity have limits.

But having come this far, we had to endure what we had gotten ourselves into and just hoped that no harm would come our way. On mid-morning of the third day, a caravan of camels arrived and we were glad to be given a ride back to El Alamein. Their leader advised us in broken English that he had been told about us by the previous caravan guide. These were reassuring words from a total stranger in the desert wilderness, and provided much comfort. Such was our experience at Ruweisat Ridge. However, we felt duty-bound to visit the museum and pay our respects at the war cemetery where soldiers of the battle were buried. A moving experience in serene surroundings and seeing ages on gravestones as young as seventeen was emotional, and every bit justified our long, arduous hike.

Our visit to Ruweisat Ridge, which we thought rigorous, paled against what these youngsters—younger than us—had endured. They boldly faced exacting circumstances, away from home and loved ones, fighting for a cause that they probably did not understand, and dying in loneliness in far-off lands.

On the tenth day, July 30, we started our trek back to Port Said to board the oil tanker for the return journey. The hike from El Alamein to Port Said was uneventful. We made the 300-kilometre trip in a day with five different rides in lorries and pickups.

Routine on this ship was similar to what it had been on the previous one, except this time, we were introduced to the ship's control room where navigation, propulsion, and communications were handled. The captain, a retired Royal Navy commodore, was welcoming and ensured us that we were well taken care of. Like the previous captain, they treated us as if we were already officers. We arrived in Rotterdam after five days, on August 5, with the highest admiration for Captain Tom Wilkins and his staff and officers.

The sea voyages exposed us to the art of leadership. Through them, we learned how a commander maintains a goal and instils morale and a hearty team spirit in the men he is leading with knowledge, inspiration, and care.

As a first experience, we witnessed firsthand that leadership as a science is not confined to the military, but is part of the interaction that takes place wherever a body of people is engaged in an activity. Such a practical experience at such a formative age was beyond anything found in books and, overall, it matured us fast.

HITCHHIKING, EUROPE

We still had another four long weeks of needing something to do. The port of Rotterdam, with its market-like scene of bustling people and congested traffic of trucks loading and unloading their cargo, meant that we had to get out and plan our next move.

We went to a youth hostel on the outskirts of the port and rested. We decided to remain in Europe, to hitchhike with no destination, and to use our Eurail pass to the fullest before it expired. The following morning, we went to the train station and studied the train schedules. We boarded a train on the nearest platform that was leaving in a few minutes. Only once we were aboard did we discover that it was heading to Dusseldorf, Hamburg, Dresden, Munich, and Freyberg.

While still in the train, we decided to go to Switzerland on sheer impulse, attracted by a colourful brochure. We changed trains to a narrow-gauge mountain train at the Swiss station of Brig. It took us to Zermatt, known for its proximity to the famous Matterhorn. Arriving at midday and walking along the high street, a "help wanted" sign posted in a restaurant window attracted our attention. This was a small restaurant that was more like a coffee bar with a counter, eight tables, and seating for thirty. The *pension* had six small suites upstairs that catered to climbers and hikers.

As I walked in, still with my backpack securely strapped to my back, an elderly, sixty-ish, bearded, cigar-chomping guy with dark glasses at the far end of a smoky corner welcomed me. He assumed that I was responding to his advertisement. He directed me to the backyard, where unskilled labour work awaited. He was too pleased to have Bill, who was waiting outside, join me, saying that with the two of us, we should be able to clean up the backyard and restack the patio tiles in one or two days, plus attend to some cleaning work in the bedrooms upstairs. With his husky voice, he offered the terms of our engagement: one hamburger, fries, and a Coke for a meal—breakfast, lunch, and dinner—and second servings at half price.

We could stay one or two nights at no charge and help out in the kitchen and clean tables. Our wages would be twenty francs each per day, and we got to keep the tips we received from serving clients. As lunchtime was approaching and we were hungry, there was no time to counter the offer: we accepted it without hesitation.

Bill and I would joke that this was our first paid job where we responded to an advertisement casually in person and were hired on the spot. We did not sit down for an interview or answer any personal questions. We had no chance to negotiate the terms of our contract, and promptly accepted the offer by simply responding, "Yes, Sir." We marveled at how fulfilling it would be if all of life evolved this way.

We happily enjoyed our free lunch and got to work. We were cleaning the backyard when a young, thirty-something, attractive brunette walked over and said in a low voice that she had placed a tray with two hamburgers and fries for us in our room. She said that we should keep it quiet, whispering that her husband was very tight and that she would be our point of contact. Whenever we were hungry, she said, we should ask her. With so much physical work, we were hungry almost all the time, and so took full advantage of her generosity.

The following morning, she asked us to hang wallpaper in two rooms. A ladder, wallpaper in rolls, glue, a tray, a brush, a sponge, and towels were on a table for us to start off. Bill had never put wallpaper on and had no clue where to start. As for me, coming from Africa, I never even knew that paper could be put on walls. I thought it had to be paint. We got out of the dilemma by asking our female friend to show us how we should start and, bingo, after a five-minute briefing, we were onto gluing wallpaper—the first time for both of us. We were overjoyed by our own success as we matched patterns with no bubbling and minimal wastage. When she came around to check about mid-afternoon, she was satisfied with the quality of our work and asked if we could stay another two or three days to finish applying wallpaper in the remaining four rooms.

We stayed there for a week and, with tips, earned over 200 francs each. We were tempted to stay on longer, but decided to move on and left Zermatt on an early-morning train on August 15 for Brig, uncertain where we were heading.

We continued travelling by Eurail for another week in Belgium and the Netherlands, each time staying at a youth hostel close to train stations and affordable on our meagre budget. We returned to Sandhurst on September 3, with still a few days in hand to prepare before our term started.

Summer break meant different things to different people. Some complained that it was too long and too expensive whilst for others it was time spent with family and nothing more. For us, it was too short. We had seen an awful lot, travelled to exotic destinations, experienced hiking and backpacking to the fullest, and came back with a wallet full of hard-earned money and stories to entertain to no end. This most joyful experience would launch us for another adventure the following year.

WESTMINSTER ABBEY, COLOUR PARTY, LONDON

A historic event awaited on December 12, 1963, when Kenya became an independent country. The first batch of three of us at Sandhurst—Jackson Munyao, Bernard Kiilu, and I—were selected to form the colour party. We were to carry the new Kenyan flag into Westminster Abbey in London as part of the celebrations to mark Kenya's independence. All the other cadets, who by now numbered about twenty from Mons, lined the red-carpeted walkway on either side of the street to the main entrance door of the Abbey.

Seating in the Abbey was full and the overflow assembled outside, spilling onto the street. Reporters, cameramen, and television crew had set up their camera stands at vantage points, anxiously waiting to capture the historic event unfolding.

Prime Minister Sir Alec Douglas Hume with his ministerial entourage arrived. The Kenyan delegation, headed by the newly appointed High Commissioner James Karanja, was welcomed by loud cheers and thunderous clapping. He was dressed in his exotic but ceremonial and elegant morning suit with tails, customary for diplomats in London. History was unfolding. Seating in the Abbey was by invitation and full. In keeping with tradition, nobody was allowed to stand unless called by the vicar at various stages of the service.

At midday when the Abbey bells tolled, we the colour party ceremoniously slow-marched into the Abbey via the main entrance, with

Jackson carrying the new tri-coloured Kenya flag and with Bernard and I as escorts. The sheer moment of entering the Abbey presented a focal point for everyone. All eyes were locked on us three. This lasted a long two minutes to a rousing standing welcome with the pianist playing soft music, stirring emotions.

The flag was handed to the vicar in a well-rehearsed military tradition. He, on receiving it, placed it on the altar. We marched off to the side and sat in designated chairs for the service. The traditional service was conducted by the vicar and was followed by speeches by Prime Minister Hume and High Commissioner Karanja. Both messages were carefully crafted. They conveyed good wishes for a new Kenya's peaceful transition to nationhood, stability for the future, and a productive and successful membership in the Commonwealth. The service lasted about forty-five minutes. At the end, we, along with members of the diplomatic corps, cabinet ministers, bureaucrats, and invited dignitaries, retired to a reception room at the Abbey for socializing.

Meanwhile, at the same time, a massive parade was held in Nairobi with Prince Phillip, the Duke of Edinburgh, representing the Queen, handing over instruments of independence to the new prime minister, Jomo Kenyatta. Kenyatta had studied anthropology at the London School of Economics and travelled widely in Europe. At seventy-three, with his signature beaded cap, flywhisk, carved cane, and goatee, he signified a majestic personality of a tribal elder and drew a widespread following from across the country. Suspecting that the successful white settler community fearing a backlash may leave en masse and cause economic meltdown, Kenyatta pacified their concerns in his address: "We are all human beings. We all make mistakes. But we can all forgive. This is what we need to learn in Kenya. Where I have harmed you, I ask forgiveness. We must put the past behind us." And, as a clarion call for communities to work together, he coined the motto, *Harambee*, the country's motto: Let's pull together.

The word *Harambee*, meaning self-help, was adopted by Kenyatta to foster community self-help and share difficulties. It is the national motto of Kenya and appears on its coat of arms. It is widely rumoured that its origins go back to when Indian labourers were working to build the Mombasa-to-Kampala railway. Kenyatta saw them carrying heavy loads of iron railing and wooden sleeper blocks, chanting *"Har Har Ambee"* in praise of the Hindu goddess deity, Ambe Mata.

Kenyatta saw the labourers communicate and work together in harmony. He saw in this a rallying call for the community to work together and, hence, decided it would be an appropriate motto for a new Kenya.

FRANCE—EXERCISE NORMANDY SCHOLAR AND MOULIN ROUGE

Part of our training at Sandhurst took us to the battlefields of D-Day landings on the beaches of Normandy, France, under Exercise Normandy Scholar. It was here that, on June 6, 1944, Allied forces mounted the largest sea-borne operation, codenamed Operation Overlord, of WW II, to liberate northwestern Europe from Nazi occupation. The object was to see firsthand the tactical-level scenarios and understand how command and leadership were exercised in taking troops to the beaches in landing craft. These were supported by naval gunships and aerial bombardment on enemy dug-in positions against stiff Nazi defences. Our weeklong base was the small village of Bayeux, which had developed skills and a reputation for making and selling tapestry. These skills were passed from generation to generation, and were the mainstay of the neighbourhood economy.

Bayeux was connected on the rail system with trains going to Paris via Caen and Cherbourg. We stayed at a hotel and the schedule was meticulously planned. We visited various battlefield sites that had been preserved and maintained for students and tourists showing where the Nazis had bunkers, tunnels, and massive concrete gun emplacements. The famous beaches of Utah, Omaha, Gold, Juno, and Sword where the Allied forces landed were also clearly sign-posted and marked.

Experienced, knowledgeable, enthusiastic guides, ever eager to relive moments of history, would give short, lively presentations on the Allied landings and how they advanced under heavy Nazi fire.

The beaches were outfitted with wooden stakes, metal tripods, barbed wire, dummy mines, and booby traps to highlight the treacherous conditions soldiers had to endure under heavy Nazi fire soon after jumping off landing craft whilst in knee-deep water and not yet on dry land.

The experience was emotional given that not many made it past just a few metres. Subsequent waves of troops were most likely affected and probably demoralized by having to jump over and past dying comrades while sensing that a similar fate awaited them. But it is amazing how a human reacts in the face of such adversity. He draws on courage and determination to overcome and keeps going, with no regard for personal safety. No wonder words such as comradeship, *esprit de corps*, courage, selflessness, and leadership had a distinct meaning in battle.

We immensely enjoyed the first three days of tours going back in history and were just as privileged to have witnessed firsthand what real soldiering meant. In between, we had two spare days for some more battlefield visits and sightseeing at our pleasure.

Seri Charusreni from Thailand and I were sharing a room at the hotel. On impulse, we decided that, having come so far and with no prospects of returning, we should ask permission to visit Paris. Our aim was to visit Moulin Rouge, the famous cabaret venue. Anticipating a refusal, we decided not to tell our instructor, Captain Inkin, and to sneak out discreetly. We only told our cadet leader, Ewen Graham, who was to join the Argyle and Sutherland Highlanders, with a soldier's promise to keep our ambitious plan a secret. We researched through brochures in the hotel lobby and asked the concierge only basic questions on train timings without alerting him to what we were up to. The same evening at supper time, we walked to the train station, two kilometres away in Bayeux, and boarded the train for Paris. We arrived three hours later at St Lazare, at 9 p.m.

Knowing that time was against us and disregarding the cost, we took a taxi for five francs and arrived in ten minutes at Moulin Rouge in the entertainment district of Montmartre. The main entrance was intimidating. Limos were parked outside and guests were arriving in formal black tie, the ladies looking glamorous in their upscale gowns, bedecked with jewellery and smiles aplenty.

The doorman was a well-dressed, tall gentleman who kept a soldierly erect posture and wore a row of medals. He looked every bit the figure of authority and in control. Nervousness was taking hold of us along with thoughts that we were foolish to have embarked on such a mission. We wore dark outdoor jackets and white shirts and carried neckties in preparation if an opportunity arose to wear

We both quickly put on a tie and hesitatingly approached the doorman. We introduced ourselves as being part of a group of cadets from Sandhurst visiting the battlefields of Normandy. We had specially come to Moulin Rouge and would be returning on the first train very early the following morning.

We requested permission to go inside for just a few minutes to see a part of the show, as we had neither the time nor the funds for the show and its expensive entrance charge. We assured him that we were not going to sit down for a meal, and only wanted to be able to say to our friends that we had been to Moulin Rouge, a sort of a checklist item we wanted to tick off.

Satisfying himself of our sincerity and perhaps sensing our youthfulness, naivety, and, probably, creativity, he said that he had never received such a request in the ten-odd years he'd been there, and asked us to wait outside, saying he would do something for us. We just had to wait. About half an hour later, he called and directed us to an empty table that still had a champagne bottle, perhaps a quarter full, and some used champagne flutes with some left over. Our spirits lifted when he told us that he acted as a guide for battlefield tours on some days in Cherbourg.

This was not the time to be socially suave and so, nervously, we quickly took our seats. It was about midnight.

The Moulin Rouge girls' performance was in progress and, much to our delight, we enjoyed the spectacle, ensuring we didn't empty our used flutes or whatever little was left in the bottle because, if we did, the waitress would clear the table. We had a feeling that the waitress had been told about us by the doorman, as she brought a platter of cheese and crackers. An hour later, it was time to leave and we had to reluctantly make our way to the train station for the early-morning milk-run train for Bayeux. Sadly, in our eagerness to exit, we did not leave a tip at the table and whatever we could leave would have been insufficient. But I gave the doorman my Sandhurst necktie which he happily took, saying that he would wear it proudly and always remember the strange circumstances of how he got it. I was just as happy to part with a memento of much significance.

We got back to Bayeux by about 0700 hrs, and took a taxi to the hotel. After a shower, we went straight to the dining room to make up for the night's hunger. Ewen Graham was overjoyed that he would not have to mark us absent and report us missing. Not a word leaked out of our exciting adventure, and we continued with battlefield presentations for another two days. On our return, we would share the tale of our discreet escapade enthusiastically and with great pride to the utter disbelief and perhaps jealousy of many. But we had mild concerns that we may face penalties for deviating from an itinerary without permission. Much later, our instructor, Captain Geoffrey Inkin, heard of what we had done. He commended us for our initiative with a warning that such a repeat act without permission would amount to disobedience and reckless disregard for authority, and would get us packing for home.

We took many risks indeed to circumvent regulations, but the intent always was for a productive and useful goal. We wanted to exploit opportunities that we assumed would be looked upon as an extension of initiative rather than a pursuit of selfish or mischievous purposes that could bring disrepute.

ICELAND, IN A FISHING TRAWLER

Sandhurst had organized a program under which several families offered to host cadets at home on Easter and Christmas breaks.

These families had close affiliations with the army, and this was their way of staying in touch. For overseas cadets, this was a fantastic opportunity to better understand and live the British culture and spend quality time in a homey environment. I was put in touch with the Low family in Scotland with whom I spent Christmas in 1963. They lived in Kirriemuir, just outside of Forfar and a short distance from Dundee.

Bill Low had built a successful business in textiles and was a well-respected figure in the Edinburgh business community. He later undertook to develop and rejuvenate the waterfront to much success. Bill was an outdoors-type guy with a family connection to the Coldstream Guards. He lived in the country and would take me pheasant hunting on the rolling hills adjacent to his vast farm. It was here that I was introduced to the world of fishing. I happily accepted an offer from his business associate, Jack Daniels, who owned several fishing vessels, to go on a fishing trawler to Iceland, starting the journey in Glasgow in summer.

I had no experience in fishing and had never been on a fishing boat. On the first day or two, I developed a liking for a traditional Scottish breakfast of kippers, a salty fish with lots of bones. At first, I found it strange and rather exotic in relation to the porridge, toast, and eggs I was used to. But as one of the pleasures of travel and discovering new cultures, it did not take long before uneasiness gave way to enjoyment of this exotic, nutritious first meal of the day.

With expectations of yet more new experiences to come, I was overjoyed at going to sea, albeit in a small boat with a small crew. The adventure was going to be nothing like the previous oil tanker experience to Egypt.

Jack had teamed me up with Peter McMillan, a University of Edinburgh student whom I had met at the Lows' family Christmas gathering and whose personality I quite liked. He liked socializing and challenged his potential to the limits and was a good mixer. Peter had attended a reservist army camp and his family also had links with the famed Coldstream Guards. Peter was under pressure from his family to serve in the Coldstream Guards so as to keep up a tradition. He was anxious to either attend an officer cadet training unit affiliated with the university or go through Eaton Hall, Mons, or Sandhurst, which was his preference.

Peter read literature and maintained that he would expand his interests to include military history and, hopefully, this would hold him well. Peter was a handsome, well-spoken youngster who played polo and tennis, and enjoyed entertaining with a guitar.

Peter was not known for an outdoors life since socializing and a vibrant party life left him with no time. Girls would gravitate toward him for affection and it was a pleasure to watch him handle such sensitive relationships with tact. On one occasion, he struggled to pacify one tricky situation where he had inadvertently invited two girls to the same party. As luck would have it, his friend's date fell sick and he was able to get out of his predicament. On several occasions, Peter found it difficult to balance schoolwork with social escapades, and he would often be reprimanded for late submission of assignments. But his engaging personality, great sense of humour, sense of creativity, and eagerness to be an active team player meant I enjoyed his company, and we got on very well. Jack Daniels had taken a keen interest in satisfying himself that we would make a good team for a trawler expedition to Iceland.

I was thrilled to receive this offer to go to Iceland for a variety of reasons. Iceland was the remotest place one could think of since not many people had visited the island, let alone knew where it was on the map or anything about its history, people, geography, culture, or lifestyle.

Iceland had a population of around 300,000, and a language of its own that its citizens preserved vigorously. The country's healthy diet of fish or lamb, Nordic lineage going back to the Vikings, lack of standing army, very few paved roads, and rural houses made of turf piled on the roof made for an exciting experience. The island sits on a volcanic plate with numerous volcanoes that erupt often, and the land is full of geysers. Until the 1950s, villages and small towns were connected with rudimentary tracks. Asphalt roads with tunnels and bridges were constructed only in the 1960s.

A recent volcanic eruption in the Atlantic Ocean in 1963 close to Iceland had attracted worldwide attention. The volcano spewed out lava in huge quantities. It piled up from the ocean floor to above sea level, thus creating an island. This was a unique event, and atlases had to be revised in response. This new island was named Surtsey. Such a development aroused interest in this, the world's newest island. There was nothing like it in the world, and it would be a unique, once-in-a-lifetime experience to see.

On July 20, at midday, we set sail in an ancient-looking trawler out of Glasgow with ten men for a two-week sailing trip to Reykjavik, Iceland. The object was to deliver some boxes that were, I believe, filled with textile rolls. On the return journey, the route took us along the Shetland Islands for actual fishing. The captain of the trawler was Mike Smith, who liked being addressed by his peers simply as MS.

He welcomed us warmly, and assured us of a good time ahead. His assistant, Charles, introduced us to other men and showed us our cabins. He walked through our routine for day and night, and explained how to respond to various drills in case of an emergency.

We knew at the outset that this was not going to be a joy ride with luxury, but a spartan two weeks of full participation in all the activities. We had to show interest and enthusiasm in whatever tasks were given us; failing that, we would be labelled a liability. This would be a poor reflection not only on Jack Daniels and the Low family, but on Sandhurst, as well.

The first evening exposed us to things to come. We had to steer the capstan on the bridge. The bridge was rudimentary: both sides were open and there was a chain across for safety and protection. The operator would stand, holding the capstan wheel with both hands. His task was to ensure that the indicator was maintained in the direction marked on the huge dial by the trawler navigator. A log sheet on the small countertop was the shift roster. This showed the time of check-in and check-out. A Plexiglas screen was our window and provided mild protection from oncoming wind. Our shift was two hours long.

During the daytime, we would operate the capstan alone; at night, we would be two. This was designed so that we would keep an eye on each other to ensure that we remained awake. The sides were kept open on purpose to ensure that the operator didn't fall asleep in the windy, rolling conditions. This was not a comfortable assignment. Peter was brought up in relative luxury and such exposure was not for him. There was no way of turning back, and two long weeks of suffering lay ahead. Peter and I started off with arguments, accusing each other of inviting the other on this arduous, inhospitable adventure. We questioned reasons why we had to accomplish any goals in such difficult surroundings and where they would lead us.

As soon as the captain or the navigator stopped by to check on us, we quietened, but resumed our quarrel as soon as they left. This was not going to be an enjoyable two weeks. Our recent friendship with hopes of enduring vanished. I had to rough it out, regardless of how I felt, because I knew that my conduct and behaviour may be reported to Sandhurst with adverse consequences for the future.

Peter was concerned about the shame he would inflict on his well-respected family. With such thoughts going through our minds, we reluctantly accommodated each other, but it was obvious that any warmth between us was disappearing. We became quiet with each other.

We were relieved after two long, agonizing hours and were invited along with four others for a drink by MS in the wardroom. Generous portions of rum broke the ice. After listening to stories of each other's backgrounds by way of introduction, the atmosphere became friendly with an assurance that we were with a welcoming team and were going to have a good time. Our argumentative discomfort from the bridge slowly gave way to accepting that we had no choice but to make the best of whatever lay ahead. The following day proved just that. We were assigned mentors and introduced to various jobs on the deck and engine room. These tasks were maintenance-related, some requiring no skill, just muscles.

Other jobs required some mild coaching, including planning a route and learning how the compass played a vital role at sea, and how communications were maintained by both voice and Morse code. We were introduced to emergency drills and safety procedures for accidents and medical emergencies, and learned, most importantly, about the use of life rafts and survival gear.

A slight uneasiness was inevitable in our cold, windy surroundings despite the waves not being too high. Even I began questioning why I had accepted to rough it out with no goal other than a novel experience for which I was not really prepared. Many thoughts would haunt me, aside from not being a swimmer: what if we jumped out into the cold water at night with life vests buckled and did not make it to the rescue raft? What about drowning? What about being eaten by sharks? What about enduring hunger and thirst? How would our families react in the warmth of their tropical surroundings?

Such thoughts are inevitable when the odds are against you and your choices for self-assurance and comfort are limited. One quickly recognizes the value of comradeship and the need to be interdependent; we simply had to be supportive of each other.

On the fifth day, we arrived at Reykjavik in the early morning and remained in port for three days. The trawler went through a routine maintenance check-up, cargo was off-loaded, and supplies of fresh water and food were loaded. We got a ride downtown by one of the truck drivers who was only too happy to practice his English with his passengers. We managed to communicate with sign language, including an arrangement for him to pick us up around five o'clock and take us back to port, a distance of some fifteen kilometres. We got off in the commercial core of Laugardalur, named after a hot spring close by where women would take their laundry to wash. We walked around and found everything very expensive and beyond affordability. Even regular snacks in cafés were over overpriced for us.

By tradition, cafés and restaurants posted their menus with prices in a picture frame outside their doors so one knew the cost before entering and thus could avoid any embarrassment later on. We just kept wandering, almost aimlessly, until it was time for the five o'clock ride back. We hoped our sign language had been well understood.

Our ride was waiting and we were back at the port in no time. There, MS was interested to find out how we had made out. Since we did not carry any bags, it looked like we had not done any shopping. He was relieved that our wallets were intact and, finding out that we had not had lunch, he immediately ordered the chef to prepare an early dinner for us. Such was the character of MS: a fatherly, supportive figure who was very much concerned about the welfare of those he worked with.

The sailing back took us along the Shetland Islands for fishing, which was carried out by casting long nets. Fishermen mechanically pulled in the catch and deposited it in the huge storage tanks on the lower deck. It did not take long for the tanks to fill, as the ocean was rich with trout. A day around the Orkney Archipelago and its 100-plus small islands took us back to Glasgow. Bill Low and his charming wife, Elizabeth, and Peter's parents, Mark and Gail, were waiting to meet us. They were overjoyed to see how satisfied and happy we looked and MS's generous compliments added to a warm welcome. That night, we were all entertained to a sumptuous dinner at home hosted by the Lows. After another few days in Kirriemuir visiting pretty small coastal fishing villages, I returned to Sandhurst in late August to prepare for our final term.

SOVEREIGN'S PARADE, SANDHURST

The final term was devoted to summarizing our two years of training. This included group problem-solving projects that highlighted teamwork and leadership, with the focus always being on keeping the soldiers' welfare and interests foremost. We were reminded repeatedly that it was our duty to get to know the men we were leading *better than their mothers knew them!*

It was only this way that the men would bring their problems to us in an expression of trust and confidence. We sat for examinations and had final reviews of our performance. Our instructors gave briefings on the immense responsibilities that awaited us on return to our regiments.

Various suggestions and advice had a common theme: to accept praise on behalf of the men as any success was due to them, but to also accept full responsibility on their behalf if things did not work out well since the men only followed what we as leaders had told them. We attended several one-on-one interviews with various faculty members, where our performance was discussed in detail with strong points emphasized and weak points that required remedial action identified.

Parade rehearsal, Sandhurst, 1963.

Our graduation culminated in the Sovereign's Parade on December 17, 1964. General Sir Roderick McLeod was the reviewing officer on behalf of the Queen. In his address, he emphasized the significance and privileges of getting a commission and the immense responsibilities that it carried. He illustrated that, for those going into the British Army, commissions would be signed by Her Majesty in her personal hand. Such was the esteem of becoming an officer. He covered leadership and officership qualities, saying that the biggest authority given to us as officers was to lead men into battle where some may get killed as a result of our orders. He said the assets entrusted to us as officers were human and we must earn their respect and confidence by example, knowledge, and integrity. It was our task to know our men and recognize their strengths and weaknesses. We had to put their welfare before ours. We had to upgrade our knowledge all the time. We had to understand the functions of staff work, command, and control and have general knowledge of tactics and weaponry to earn their trust. They were fine words that adequately summarized what we had learned over two years.

In keeping with tradition, the colours were trooped and the parade marched past him, giving him an eyes right. We in the senior division entered the old building via the grand entrance, marching up the famous steps, with the adjutant following us, also via the grand entrance, on his white horseback.

A celebratory lunch awaited us in the dining room, along with bear hugs and handshakes to wish each other success for the future. In the evening, we went to a well-attended celebratory commissioning ball in the academy dining room.

We proudly showed off our "pips"—stars on our mess kit epaulettes, for the first time at midnight. We were now fully fledged second lieutenants at the very bottom of a career ladder with ambitions to reach for the top, one day. Sandhurst had prepared and launched us with a strong foundation, and it was up to each and every one of us to make the most of what we had learnt and to *shape our destiny*.

RETURN TO KENYA

After graduation in December 1964, I attended a short, two-month small arms course at the School of Infantry in Warminster, in January of 1965. This refreshed our minds to platoon tactics using a standard 7.62 mm assault rifle, a Sten gun, a Bren gun, a 9 mm pistol, 36 mil hand grenades, and 2" and 81 mm mortars. We were shown how to lay mines and booby traps, and how to discover them in hostile surroundings where the enemy had laid them. Bigger challenges awaited back in Kenya as a breakaway insurgency in the Northern Frontier District, the NFD, was gaining momentum as neighbouring Somalia supported it and laid claim to the region.

Firstly, a warm welcome awaited at the waterworks camp in Nairobi by the British commander of the new Kenya Army, Brigadier John Hardy, who introduced the recently retired commander of 70th Brigade, the KAR, Brigadier Miles Fitzalan Howard, for a few words. Miles was later to inherit the dukedom of Norfolk. After the customary congratulations and good wishes for success for the years ahead, he said a few words in closing that did not make sense at the time—or perhaps we were too naïve to understand. The underlying depth of the message resonated with us all years later: "Gentlemen, planning for your retirement begins as of today. I have instructed the paymaster to deduct twenty percent of your salary at source and this will be automatically deposited into your savings account at the post office. If you do not wish to have these deductions made, you are free to do whatever you want."

None of us had the courage or desire to challenge a senior officer whom we held in very high esteem, so we reluctantly kept quiet and accepted the advice, much like an order not to be disobeyed. With no housing, food, transportation, clothing, or medical costs, our pay was really pocket money for cigarettes and beer.

Years later when the time came to buy a house, this advice was relived with much happiness. We had all accumulated a pile of money and were to buy our first houses with substantial down payments.

THE SHIFTA INSURGENCY

The security situation in the NFD was worsening, with Somalia maintaining that the region had been illegally taken out of greater Somalia by the colonialists. Indeed, NFD came into being in 1925 as it was carved out of Jubaland from present-day Somalia. The district, under British colonial rule, had a population of Somali ethnicity that was closer to Somalia in its cultural make-up. The northern part of Somalia was ceded to Italy as a reward for Italy's support of the Allies in WW I. Britain retained control of the southern half of the territory, later called the NFD, and gave responsibility for its administration to Kenya in June 1960, just before Somalia got its independence. This was the background to what was to follow.

This insurgency was called the "Shifta insurgency," a secessionist movement with a small following in 1960. It became active and gained momentum in 1963 in the NFD at Kenya's independence with the goal of joining their fellows in a greater Somalia. Shifta is a Somali word for banditry. It took the form of banditry with random attacks on villagers, planting hidden roadside bombs, carrying out ambushes, killing innocent people in marketplaces, and creating fear in the population. The young Kenya Army, growing but still in its infancy, had the task of containing this volatile situation. It was supported by a very young Kenya Air Force, the paramilitary General Service Unit, and the Kenya Police.

Intelligence-gathering was a critical factor in operational planning. Special Branch headed by the renowned James Kanyotu, Kenya's J.

Edgar Hoover, did a remarkable job in providing such intelligence on a daily basis. This allowed various security committees to evaluate and plan operations. These security committees were at the district, provincial, and national levels, chaired by the district commissioner, the provincial commissioner, and the under secretary in the ministry of internal security. Meetings were held weekly.

The security situation in the early 1960s had deteriorated and required armed escorts for food and other convoys resupplying commercial enterprises in the NFD. Travelling alone was inadvisable as it invariably led to being ambushed.

DEPLOYMENT IN THE NFD

Soon after returning from Sandhurst as a newly commissioned officer in late February 1965, I was posted to the newly formed 1st Battalion Kenya Army as a second lieutenant, based in Nanyuki, on the foothills of Mount Kenya. The unit was renamed from the remnants of the 11th Battalion, KAR, who, following independence, *Uhuru,* had mutinied and was disbanded. After the mandatory two-week leave, I was sent with my 3rd platoon, A Company, to our company based in Isiolo. From there, after an orientation of two weeks, I was sent to Merti, an outpost trading centre some 500 kilometres from Nairobi. This was almost in the middle of nowhere in the vast expanse of NFD inhabited by nomadic tribes, constantly on the move in search of water and grass for their cattle. Roads that were more like camel tracks formed a haphazard network linking villages with the main road going parallel with the River Tana.

As a twenty-one-year-old, I had around 100 soldiers, as opposed to the regulatory thirty-five. My mandate was to be on detachment for six months and, for this long duration, I was supplemented with additional support and administrative personnel. We set up camp consisting of 160-pounder single-post tents, outside of the Merti trading centre. The site was close to a rough grassy but flat ground that had been used as an airstrip a long time before. We started planning our routine. This was going to be home and we had to be content with making the most of whatever army-issued supplies we had for our needs and survival: field rations, fuel, tents, medical supplies, and additional stocks of mortars, hand grenades, and ammunition.

What I was not going to do was sit out the six-month deployment waiting to be attacked or ambushed. The dullness of a base with nothing substantive to do can quickly lead to rumour-mongering, loss of confidence, lowering of morale, breakdown of discipline, mistrust in comradeship, and bickering.

We had to be creative in making our living comfortable, and this meant building a relationship of some kind with the locals.

We had to win their hearts and minds and assure them that we were with them and not against them. This was not an easy task, given language and cultural barriers and the inevitable fear that we were disturbing their free-roaming nomadic lifestyle. We were intruders in their territory, and so had to assure them that we were there to stay only for a short period to enforce security for their benefit.

Our communication link with company and battalion headquarters was by vintage, vehicle-mounted radios. We were required to send a daily situation report, *a sitrep*, every evening summarizing the events of the day. Messages were brief as they had to be manually encrypted into *Slidex* code, time-consuming work. Further, the atmospheric interference also meant that even Morse code transmission could be uncertain.

Added to this was the perpetual breakdown of our equipment and every effort had to be made to pass on the key elements of a brief report as a priority and fast. Complaining about poor equipment was no solution as we all knew what was available. We just had to make the most of what we had.

A well-trained medic, Corporal Kimani, who was conversant with various field medical needs, was assigned to us. He had an assistant, Private Jonathan Kibwego, who had rudimentary knowledge in health care but abundant enthusiasm to become a fully trained medical orderly one day. They carried basic medical supplies consisting of painkillers, morphine, quinine, diarrhea-suppression tablets, water-purification tablets, an assortment of bandages to treat minor wounds, a few braces to support fractured limbs, anti-snake-bite serums, and antibiotics that were administered by injections.

Our field rations for the soldiers were a mix of canned meat, powdered milk, powdered mashed potatoes, sugar, salt, maize meal, dried peas, kidney beans, hard tea biscuits, and tea leaves.

Where possible, we could purchase fresh meat and vegetables locally. The officers had a special ration of canned sardines, bully beef, canned baked beans, tea, and hard tea biscuits. I realized that the ratio of cooks to men was not right: two cooks for 100 men and one cook for one officer. I was in a dilemma. We had just inherited customs and protocol from the British, whose concept of leadership and man-management was colonial. Distance had to be maintained between the rulers and the ruled, minimizing social interaction. This was a key element in empire-building with the justification that it was necessary to maintain divisions to avoid opposition to rule and gain widespread cohesive support.

This is not what I had learned at Sandhurst where morale, man-management, and discipline were meant to be key elements of fostering trust, confidence, and loyalty in men who were expected to show exemplary conduct. I would see two cooks laboriously work almost all day to feed three healthy meals to 100-plus hungry, tired men who had returned from long patrols carrying heavy loads of ammunition, mortars, water, and other supplies. The soldiers would line up, holding their mess tins by the cooking pot. Two cooks would cheerfully serve them generous portions. And I would sit on a chair at a distance served by one cook using plates, knife, fork, and spoon, and with my batman on standby to run any errands.

On the third day, with the full knowledge that I was conducting a breach of an established routine, I called in CSM Nduati, the company sergeant major, and told him that my cook was not going to cook for me any more and that he was to join the other cooks forthwith. I also said that I was no longer to be served separately, and that I would eat the same meals that my soldiers were served in a mess tin and would be the last, after everyone else had been served.

This one gesture brought me very close to my soldiers who shared their daily living routine with me and who saw nothing different in me except that I was in command and in control of their lives.

Bigger tasks lay ahead. We had to concentrate on why we were in the field and what mission we had to accomplish.

A young twenty-one-year-old second lieutenant in his first command of over 100 experienced men, some twice his age, just three months after graduating, was a challenge in itself. Managing people in field on active duty poses unimaginable challenges to morale, discipline, and leadership. These subjects had formed the core of our two-year learning at Sandhurst where we were constantly and repeatedly reminded that we were dealing with men and that we should always put their interests before ours.

Now, facing a hidden enemy around us and not knowing when we would see action, there was no time to prophesize; action had to be taken immediately. Priorities had to be identified and set. Our resources were limited. In my detachment, we had only four three-ton trucks and an FFR (Fitted For Radio) Land Rover. This would be my mobile command when on the move.

After just three days of settling in and going through a rudimentary daily routine, I knew that a mammoth task lay ahead. First and most importantly, I had to be accepted by my soldiers willingly as an effective team player who did more than simply exercise his authority over them and pull rank. I discovered that I had a mixed bag of personalities in my command: some obedient to a fault but who would wait out to be told what to do, some smart alecks who knew everything and always had an opinion on why things should be done differently. And there were some whose courage, strength of character, and enthusiasm showed that they were going to be my backbone in whatever I wanted to do.

The challenge now was how to mould this body of over 100 men into a strong team with high discipline, morale, team spirit, and willingness to execute our mission to the best of our ability. The soldiers who had been drawn in from other battalions had not yet developed the regimental spirit of a new unit. A common thread with all was an element of fear—not uncommon with a soldier launched into a combat role and having to face life-and-death encounters every minute.

This was not a training exercise. This was reality and, as a platoon commander, I often thought of the pain and sorrow of losing a soldier in a fire exchange as a direct result of my plans and orders. I quickly planned a strategy on how I was going to conduct myself within the vast terms of my mission. I was given a free hand to do whatever I wanted in the execution of the plans for us to patrol, seek out, and destroy the enemy. I was overwhelmed by such authority and a little uncertain if I could deliver upon expectations.

Language was a subject to be addressed fast. Although many spoke and understood English, they were naturally comfortable in Swahili, as that was the bonding language of folks beyond their tribal dialects.

My knowledge of Swahili was rudimentary and it often triggered amusement as a straight translation from English was effective only because the soldiers were too understanding and acknowledged my efforts. However, communication being an essential component of a relationship, I immersed myself in learning Swahili as fast and as best as I could from my soldiers. They were most helpful, and felt proud to be my teachers. I soon found out that Swahili is a rich language and communications with the soldier, the *askari,* became easier and respectful. The grammar may have been faulty, but the soldiers showed an ability to make sense of what was said, thus displaying trust and integrity in the overall well-being of the team. The key was to do the right thing rather than play on words and detect wiggle room to get out. It was only a short time before my Swahili had progressed impressively to a level at which I could comfortably rate it as my first language.

My priority was to build a strong team with high morale, discipline, and enthusiasm to execute our mission to the best of our ability. This was my family for the next six months.

I had to be accepted by them. They had been placed under my command and, as such, I was responsible for all their needs, demands, hopes, and aspirations. Of Kenya's forty-two tribes, I had a representation of over twenty-five, and I had to know each and every one of them by name, rank, background, family, home village, strengths, weaknesses, loyalty, and competence. As with any body of men, I had to identify future non-commissioned officers and those with the potential to attend a course to upgrade their skills. It was a tall order for a young, inexperienced, but ambitious officer. A deep knowledge of the inner workings of the tribal structure was needed as language, customs, and traditions were so different and yet so interdependent on each other. A good deal of commonality had to be fostered.

As in any organization, there are those who would follow orders promptly and those who have a good reason to be different, to say they couldn't go on a patrol because of a tummy ache, headache, or any of those hidden handicaps that require a professional medical examination. I would have none of that, but instilling discipline with threats and recourse to punishment was not the answer. I was new to them, and they could easily lead me astray if I became overbearing in my efforts to maintain my authority. I had to be firm, but fair. We were in battlefield conditions, and discipline had to be instilled by example, loyalty, trust, and confidence.

A multitude of important tasks lay ahead, and the difficulty was prioritizing them. The most important one was to be accepted by the men who would willingly submit to my authority. Team spirit was there, but it had to be boosted up as the newly formed 1st Battalion was made up of soldiers from other battalions who had not yet fostered a closeness with each other.

Being launched into a combat role from the very recent celebratory mood of having gained independence was demanding on nerves.

Families were left behind and it was natural for each and every one to worry about something back home, including the health and welfare of their families, their livestock, personal issues, etc. Some who had experienced the Mau Mau emergency were of great assistance in helping cushion weak minds.

Physical fitness was going to be key to our long-distance aggressive patrolling schedules as they were done on foot across hilly, rocky, inhospitable terrain. Building endurance in sweltering heat across desert-like conditions would form the bulk of our routine.

Aggressive patrolling would also bond us all into a team and give us the initiative over the enemy that would convince the men that we were on the offensive and hunting the enemy and not the other way around.

I decided to split my men into four groups of roughly twenty-five each. Each group was to be led by me, my platoon sergeant, Sergeant Ndegwa, and section leaders, Corporals Nyamweso and Onyango. Patrolling was to be daily, leaving the base at different times but usually around midnight or well before dawn to take advantage of cool weather to cover distance. The three remaining groups at the base would practice weapons drills, and attend lectures on laying ambushes, administering first aid, gathering intelligence, and other general subjects. Every soldier got involved, and it was encouraging to see a warm, cordial atmosphere prevail. Each soldier introduced himself in depth about his background, tribal structure, village life, family life, and other general matters. Such talks in small groups and on a volunteer-attendance basis were very effective in not only familiarizing us with the depths of each tribes' values, but in bonding us all together as a team. It enhanced morale and made us interdependent.

Corporal Kimani, our medical orderly and *muganga,* the doctor, would fascinate us with his knowledge of traditional medicine.

He would go beyond army-issued medications and resort to the traditional care available from plants whose juices and mixtures worked wonders. He had a solution for any medical problem, be it diarrhea, headaches, aches and pains, wounds, snake bites, and so on.

Corporal Oloo was the undisputed teacher. He would conduct English classes. With limited notebooks and pencils, his teachings were oral, and drew much enthusiasm. He had a flair for combining humour with instruction, oftentimes mixing English with Swahili and the odd word of Luo. He would add spark to many a dull day and was more of an entertainer than a teacher with his skill at captivating an audience. Singlehandedly, he lifted everyone's spirits enormously.

We kept away from any discussions on politics as that was not allowed. However, every soldier had to give a small talk about himself or his tribe on any subject. This way, nobody felt left out and the shy ones were slowly given the assurance of support. This built up their confidence and fostered comradeship.

Playing cards was a popular activity and much encouraged. This had to be watched closely to ensure that it did not lead to gambling, a concern we had been made aware of. Lunch was at noon, followed by full rest and sleep till late afternoon. Midday was brutally hot and it was best to sleep it out.

After dinner, served at 1630 hrs, we would have a *ngoma*—a little folklore singing and storytelling around a fire.

It is amazing how this social interaction brought us all close to each other and allowed us to know each others' backgrounds so well. A section was always on guard duty, manning various posts around the camp perimeter. As part of our operations, we would set night ambushes if the intelligence we received appeared authentic.

Oftentimes, the intelligence was misleading, resulting in disappointment. But it was never a waste of time as we looked at such deployments as practice drills.

The four groups of my platoon would take turns in aggressive patrolling with two days in a week to rest, during which time we would service our weapons. With this physical fitness taken care of, morale, discipline, and man-management had to be addressed—and addressed fast. There was no time to guess and speculate.

Morale-building, instilling discipline and a sound approach to building trust, confidence, and a strong team were key elements to focus on. I adopted the rule that I had to be able to do whatever I asked of my soldiers. By doing so, I would gain their trust and confidence. Hence, my decision to split my platoon group into four smaller units. It did not take long before the soldiers saw me as one of them—bearing similar backpack loads on patrol, carrying my own supplies, and maintaining focus on the time and route of our mission.

We had to build high morale and convince everyone about our purpose there. We humans are often scared in hostile surroundings when we don't know when and where the enemy will strike. It is very easy to take the easy route of being defensive and hoping nothing catastrophic will happen. Such a mindset, however, leads to being scared, remaining on the defensive and, often, a sense of inferiority that the enemy is superior and stronger, which impacts negatively on morale. This was not the right approach. I immediately decided that, in order to build morale and make everyone have confidence in our capabilities to succeed in our mission, we had to take the initiative and dominate.

We had to show a strong willpower; to go on the offensive and hunt the enemy. We had to find where he was, what his resources were, who was helping him, and what his intentions were. It was a tall order for a small group given the task of ensuring security in a desolate, sparsely populated area of over 2,500 square kilometres. As an indirect bonus, this gave us a free hand with minimal oversight on what and how we were conducting ourselves. All the aspects of the Geneva Convention listing the dos and don'ts were followed, but remained distant, much to our relief.

Our major handicap was getting reliable, timely intelligence upon which we might plan our operations.

We got intelligence reports from company and battalion headquarters some 300 kilometres away, and thus often late and inaccurate. I adopted a quiet strategy of building a rapport with the local nomadic herdsmen. We would provide them medical support for their families and, oftentimes, some food with the full knowledge that such a policy was against our mission directive and in breach of our terms of engagement. I took a chance, putting my career at stake, and initiated contact with the nomadic herdsmen who got us closer to their pastoral living and humble rudimentary dwellings.

We gained their trust by using our medical orderlies to provide treatment to the sick in their families. My medical orderly, Corporal Kimani, quickly gained recognition as "Mr. Doctor"—"*mbwana muganga*"—by the villagers.

Winning hearts and minds came at a negligible cost and with huge rewards. We got much more in return than delivering acts of humanity to deserving, innocent individuals. As in any insurrection and volatile situation, the local inhabitants become a pawn; they cannot avoid helping the terrorists at night by providing shelter, food, medical care, and, most importantly, the location and intentions of security forces. At daytime, they would submit to the law of the land and appear loyally abiding. It was a tough call and difficult to balance conflicting loyalties.

Going on daily patrols built our endurance and bonded us into strong teams. Through them, we quickly recognized the doers, followers, malingerers, and perpetual complainers for whom nothing went right. It was a challenge to build their confidence, gain their trust, and make them feel that they were an important part of our collective effort. These bonds strengthened and, on the third week, most of the malingerers and complainers had been reformed. They had no alternative but to give in to the overall group values.

Discipline was enforced quietly and by sound reasoning as an essential component of our daily routine. Reveille would be at 0500 hrs followed by a daily dawn inspection, brief informative lectures on weapon-training drills, and the constant practice of various tactical drills for those staying in base.

Those going on a patrol would leave base around 0200 hrs, arriving three to four hours later at an area to be patrolled or where an ambush was to be laid. It was critical to develop close bonds with the local herdsmen as they roamed the area all the time and were knowledgeable about the movements of the Shifta. We needed this information to assist in our operational planning and could not afford to distance ourselves from them. They had to believe that we were there to support them in their way of life. We also developed close contact with the small kiosk- and store-keepers in the village who were most likely to be in contact with the Shifta. We recognized that these herdsmen would also be passing information about our movements to the Shifta to whom they would be just as loyal.

In just three weeks, with such a densely packed routine, we developed close bonds with each other. Oftentimes, volunteers from other patrol groups I had formed would plead to go on patrols. My preliminary mission had been accomplished. I had a strong, loyal team of enthusiastic soldiers who were determined to maintain the initiative and seek out the enemy. Malingerers, complainers, and those with uneasy nerves were handled by their peers and gave in to team spirit. Morale blossomed. Discipline was never an issue. Comradeship grew. Number 3 platoon, A Company, 1st Battalion was in high spirits and itching for action.

I knew we were ready for combat and I had full confidence that each and every man was there of his own choice. I never noticed any signs of cowardly behaviour or attempts to remain away from going on patrols. Our biggest handicap was that we lacked good, reliable, timely intelligence. We acted on a few tips, but soon discovered that they were days old. But it was certain that the enemy was around us, waiting for his kill. We had to maintain the initiative and remain on the offensive. This meant keeping a demanding, aggressive long-distance patrolling schedule to dominate the area around us—a vast landscape almost as far as the eye could see.

OPERATION MERTI

On April 18, 1965, we got a tip from a herdsman that a large number of Shifta were going to assemble in a hilly area, some ten kilometres from our base. They were going to stockpile land mines and ammunition, and regroup to launch an ambush on a resupply convoy that was arriving to replenish our food and medical supplies. At first, I brushed off this tip as being too vague and far away from the road. Why would the Shifta assemble ten kilometres from its intended ambush site? However, on April 19, two herdsmen independently corroborated the earlier tip; there was indeed going to be an ambush and the Shifta were assembling to set up two ambushes, the second one being for the rescue team that would inevitably arrive. Now there was no time to waste. We had to act fast and ensure surprise.

I doubled my patrolling group and, with fifty soldiers, grouped them into five teams of ten men. We would drive toward the hilly area, leaving base at 0400 hrs so as to arrive at the suspected site by daybreak, then de-bus, leaving one group to guard the vehicles and encircle the enemy site with two groups, with the third group remaining some fifty metres behind in reserve.

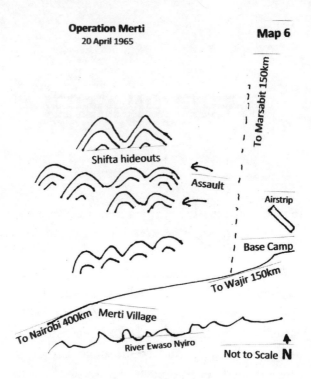

Operation Merti, Kenya, 20 Apr 1965.

As soon as we got to where we were going to de-bus, Private Jonathan alerted us to the smell of smoke from the hills, indicating that there were people there—exactly where we were heading. The plan was for two groups to go in a flanking move with a wide detour away from our vehicles to surprise whoever was there—hopefully the Shifta. Sure enough, the Shifta were indeed there, and their lookouts must have heard the rumbling engine noise of our vehicles. Just about when we were going to de-bus, we came under heavy small-arms fire from high ground ahead of us. Bullets rained on us from all around. We quickly de-bused, took cover behind small mounds, and returned fire.

Until then, I had been standing in the cabin of my three-tonner truck observing from the hatch, seconds from jumping out. I noticed that my bum was wet and as I put my hand there, saw blood oozing out. I had been shot. Moments before jumping out, I saw a fez-capped

SHAPING DESTINY

Shifta, well hidden in grassy cover, pointing his rifle at me. Seconds mattered with all the advantages of cover for him whereas I was in the open, exposed, and a moving target. I quickly aimed at him and fired a shot that killed him.

Now on the ground but still exercising control by voice, my concern was to control our ammunition stocks, since there was no back-up plan for resupply.

As firing lessened and it was daybreak, I checked on casualties and, much to my comfort, the three group leaders shouted back to confirm that they were OK and had no casualties but were low on ammunition. By 0630 when the firing had died out, we assumed that the Shifta party had been neutralized and either had been killed or that some survivors had disappeared into the hills and were on the run. We had successfully engaged them on their ground with me being the only casualty. We rounded up our five groups and conducted a physical check with the assumption all along that we may be counter-attacked.

We did a thorough search of the area and counted a good number killed. We confiscated a large stockpile of land mines, grenades, mortars, and small arms. We had been successful in our first major combat engagement within a month of arriving in a most troubled part of the NFD. All that rigorous training, unending practice drills, comradeship, and discipline came in handy and were the reason why we were so successful. This was a fitting rite of passage into the real profession of arms.

My field radio, Racal A906, was a weighty piece of equipment to cart around. It rarely worked and, when it did, it did so only briefly. We soon developed an instinct to send messages in brief, as soon as a link was established, knowing that we had just a few precious seconds to do so before getting cut off.

One of the few times we got the A906 working was, luckily, when we had our combat engagement. Luck? This time, we not only got through to our company headquarters in Isiolo, but to battalion headquarters in Nanyuki and, most surprisingly, to army headquarters in Nairobi, some 500 kilometres away. In the excitement of our engagement and taking full advantage of our wireless connection, Private Ngugi sent the message, "Sunray KIA (killed in action)." He had seen me losing blood heavily and sensed that I was near death. Sure enough, army procedures got my message passed onto channels requiring burial action.

Luckily, a police aircraft, a Cessna, was in the vicinity and, upon intercepting this message, circled overhead and landed on a flat grassy surface that at one time had been used as an airstrip by our camp. The pilot, James Callaghan, knew the area well. He was ready to take a "dead body" back.

James was pleasantly surprised to see me very much alive. After fastening my stretcher in the aircraft galley, we took off for Nairobi. My ride, from the engagement site to base, lying on a stretcher in the back of a three-tonner truck over rocky surfaces, was slow and painstaking. Corporal Kimani, my medical orderly, administered painkillers and gave me morphine to numb the wound. I recall hearing him pass a message to police headquarters that I was seriously wounded and bleeding profusely, and asking for an ambulance to meet us on our arrival at Wilson Airport on the outskirts of Nairobi. Lieutenant Colonel (Dr.) Brendan O'Duffy, our senior army doctor, met us, assured me that all would be well, and I was onto Nairobi Hospital for surgery to remove shrapnel from my thigh. A two-week stay in hospital and a further two-month recuperation at Waterworks Camp walking with assistance of a cane reminded me of my soldiers, who had become my family. I was itching to go back to find out how they were doing.

I recalled with fond memories Second Lieutenant Frank Nthiga who had come to resupply me only a week earlier to wish us success for combat. We had met outside the camp and his preparation of a roast chicken for lunch was a pleasant and welcome break from cookhouse cuisine.

KNOWING YOUR MEN

A Kenyan soldier has an inborn flair for the outdoors. He is a natural tracker, has a finite ability to sense danger (primarily from wild animals), and is loyal beyond definition. His creativity and sense of living off the ground are legendary.

During my deployment, I not only undertook to improve my Swahili, but wanted to know everything about the men I was leading. Having settled to reinforce the military side of rudimentary requirements such as morale, discipline, comradeship, loyalty, and so on, I set about now to find out more about the tribal structure with its many languages, customs, traditions, beliefs, and lifestyles.

I wanted to know more than what I knew from casual encounters at home or from infrequent discussions on the subject—which, regrettably, were never focussed on the tribal make-up but a discussion that would often be brushed off with casual, vague references to some distant beliefs.

It was ironic that we made home in a new country to better ourselves and yet made very little effort to understand the make-up of the new indigenous society.

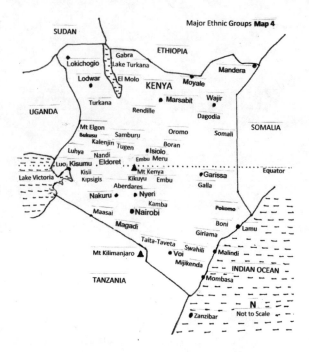

Major ethnic groups.

Kikuyu

First and foremost, I started with my platoon sergeant, Nicodemus Ndegwa, who was Kikuyu, the largest tribe in Kenya, numbering about 10 million. Kikuyu is a derived name from Gikuyu, a fig tree commonly found in Kenya called *mugumo*.

The Kikuyu worship *mugumo* and consider themselves children of the tree. They believe in *Ngai*, the omnipresent god who is also a companion to *mugumo*. According to Kikuyu folklore, *Ngai* lives in clouds, heavens, and sun, and his resting place is Mount Kenya. The Kikuyus' main diet is irio: mashed corn with beans and peas served with roasted goat, beef, or chicken and mashed spinach or green leaves, commonly called *sukuma weeki* (so named because it is readily available and affordable to see one through a week). The Kikuyu perform rituals and rites at birth and death under the sacred mugumo tree. They live around Mount Kenya, and are farmers and shrewd businessmen. They are closely related to neighbouring tribes, the Embu and Meru.

With intermarriage, the Kikuyu expanded their influence to other tribal areas. The Kikuyu are not a warlike people, but when the Europeans expanded their farming interests during the colonial era starting in early 1900s, they encroached into the rich farm lands of the Aberdares and Mount Kenya, the traditional Kikuyu heartland. The Kikuyu were displaced from their farmsteads by the disbanded soldiers of Britain who came in large numbers as settlers after the First World War and later after the Second World War. This land grab caused resentment and animosity and eventually led to the Mau Mau rebellion to oust the British.

My Kikuyu corporal, James Kinyanjui, would recall his mother giving him a background on the Mau Mau. He was aware that the essence of oathing was to keep it a secret and that, once an oath was taken, one was sworn to secrecy, the betrayal of which was death by *panga*. *Panga* is a broad-blade machete, commonly used to hack tall grass and dense bushes to carve out paths. However, the country had gained independence and times were changing. The youth were influenced by Christian teaching in missionary schools and exposed to a Western influence. Living in an urban society, the youth slowly distanced themselves from the traditional grip of their elders and were now looking at a future in cities, which meant acquiring education, skills, and the opportunity to work in urban surroundings.

For those who took jobs in the government and enlisted in the military, daily routine and demanding work schedules brought them into closer contact with other tribes and nationalities. With a sense of belonging to a newly emerging society, they felt free to discuss the key role that Mau Mau played and did not fear discussing its many rituals. With this comfort, James spoke freely about what he would hear from his mother and other elders at home. Mau Mau was an ancient secret society in the Kikuyu tribe and the name was a reverse or derivative of Uma Uma, meaning "Get out, get out." The simpler and easily understandable convenient Swahili interpretation was *"Mazungu Ayende Ulaya,"* or "Europeans, go back to Europe" or *"Mwafrika Apate Uhuru,"* or "Africans get independence."

James would continue: The Mau Mau instilled fear amongst their members who were forcibly enlisted into their ranks by administering midnight rituals and blood oaths to honour their tribal ancestors.

The Mau Mau would drag men from their beds at night and, after threats and beatings, administer blood oaths that required drinking blood, eating human flesh, eating the flesh of dead corpses, living with animals, and promising to maim or kill whoever opposed the Mau Mau, even if it was the person's father, mother, brother, sister, or any other family member.

Such oathing ceremonies that had their oaths developed from the traditional ceremonial types with brutality, sadism, and sexual overtones bordered on cannibalism, and affected the psychology of the person, turning him into a different sub-human individual. This cycle of loyalty to the movement would be repeated again and again to reinforce loyalty. Slowly, such oathing extended to other tribes, women, and children to expand the movements' base.

James would narrate the allegiance committed when taking an oath: not to sell or dispose of land to foreigners, not to be a whistle blower, not to accept bribes or be a mole, to keep the oathing secret from everyone including close family, and to worship Jomo Kenyatta as the only leader of the Kikuyu and Kenya. The penalty for failing to observe this code was death with no recourse for appeal or forgiveness. Death would come with the head being chopped off with a *panga*, the eyes would be gorged out and blood as it oozed, drunk. The decapitated head would be put on a stake in a visible location as a deterrent to others who went afoul.

Oathing was of two types. The first oathing ceremony was called the *githaka* oath, or the forest oath. This was administered by gang leaders to those who lived and operated in the Mount Kenya forest and the Aberdare mountains. The second version was called the *batuni* oath, meaning a platoon oath.

The word *batuni* is probably a derivative of the word platoon, which in Swahili was often pronounced, *platooni*. This was administered to the foot soldiers of the Mau Mau movement, soldiers whose job was to execute the orders of the Kikuyu hierarchy whenever and wherever they were sent to kill, maim, or steal weapons or cattle. No one was spared, even if the order was to kill a father, mother, or sibling.

A *batuni* oath was administered at night with the recruit taking off his clothes and lying on the ground naked, surrounded by senior Mau Mau elders. A freshly slaughtered ram would be placed next to the recruit. The ram's flesh and penis would be placed on the recruit's chest. He would hold the ram's penis with his left hand and eat a small part of the flesh with his right hand.

Often, a menstruating female, also naked, would stand close by. The elders would insert part of the flesh in her vagina and give it to the recruit to eat, cementing an irreversible firm bond. Once a *batuni* oath was administered, the recruit would be classified as a full, loyal member of Mau Mau, prepared to execute any order and authorized to administer an oath on other new recruits.

Those Kikuyu who refused to take the oath would have their ears chopped off for having refused to uphold the Mau Mau cause and, in turn, having demonstrated that they were loyalists to the government.

It is widely believed amongst the Kikuyu that oathing took on a widespread appeal in 1943 as a measure to build a secure, secret base for a rebellion in the small village of Olenguruone in neighbouring Maasailand, where over 10,000 displaced Kikuyu, now called squatters, had been relocated.

Kamba

The second person I got to know well was my section corporal, Matthew Nduati, a Kamba. The Kamba are a warrior tribe, living in dry pastoral land around Machakos, Kitui, and Makueni. They number five million and represent about fifteen percent of Kenya's population. By tradition, they are pastoralists who have slowly moved into farming. The Kamba were the backbone of the British Army's King's African Rifles. They are natural fighters; loyalists with a deep sense of duty who took great pride in being soldiers. They are reliable, brave, and ferocious with a natural flair for combat. They are pastoralists who raise cattle, are traders, and have slowly engaged in farming.

Lack of rain is often a big problem. This leads to a ritualistic rainmaking dance called *Kilumi*, a healing ritual accompanied by prayers offered to restore an environmental balance. Specific dance movements evoke a sense of spiritual connection to ancestors, to ask their forgiveness for wrongdoing, and seek their intervention for assistance.

They believe in an omnipresent god called *Ngai* or *Mulungu*, with their ancestors providing a link remembered at family rituals. In the Kamba family, a man's role is an economic one: trading, cattle herding, and farming. The woman's role is running the household and raising children. Nephews, nieces, and cousins enjoy an equal standing with one another and freely move from one household to another, always enjoying the parental closeness of a mother. The naming of children in the Kamba family has a special significance. The first two boys and two girls are named after the paternal grandparents on both sides.

Luo

Private Oloo, a Luo, hailed from Kisumu on the shores of Lake Victoria. The Luos, at around fifteen percent and numbering over five million, are the third-largest tribe. Oloo was studious, spoke good English, and would talk about how Christianity had influenced his village. The Luos did not lose land to the colonialists as the Kikuyu did, but adopted Western-style clothing and excelled in education. Oloo's folks were torn between traditional religious beliefs and a new religion, Legio Maria, a mixture of Christian, Islamic, and traditional beliefs, was spreading fast, attracting a large following. Boys did not go through circumcision rights in their early years as with other tribes, but instead had six lower teeth removed, signifying a right of passage from boyhood to a junior adult. Luos practiced polygamy and were allowed to have up to five wives, with the groom paying a cash amount and a dowry as negotiated among the elders to the brides' parents.

Their main occupation was fishing, farming, and raising cattle. A Luo diet is mainly fish, since they live on the shores of Lake Victoria. They straddle across into northern Uganda, western Tanzania, and southern Sudan.

Maasai

Private Ole Sainapu was a Maasai and a gifted tracker. We relied on his skills to guide us on patrols. Maps were alien to him. The Maasai are nomadic pastoralists, straddling southern Kenya and northern Tanzania. They number under a million, representing about three percent of Kenya's population.

Their stronghold now is around Narok and Kajiado, straddling across into Tanzania around the world-famous Ngorongoro Crater.

Their migration from northern Kenya, intermarrying, and assimilation of local tribes produced descendants called the Kalenjin and the Samburu, who now live around Eldoret and in the arid north-central regions around Marsabit and Maralal.

The Maasai believe that all the cattle in the world belong to them and that they own grazing lands and watering holes for their cattle as communal property. According to folklore, no one person can claim ownership of land.

Cattle belonging to neighbouring tribes is often a target for the Maasai looking to retake what they believe should belong to them—hence, the periodic cattle-rustling wars. The Maasai maintain a distinct hierarchical system of managing their affairs: a court of elders is the seniormost governmental body. It can issue edicts for settlement of disputes, often in terms of cattle to be handed over by the guilty party, or the offering of a daughter to make peace with a warring faction.

The end of life for Maasai has no ceremony and folklore has it that when one knows that he is nearing death, he or she will say their good-byes and walk into the wilderness alone and without a spear or shield to defend themselves against attacks from wild animals like lions or hyenas. Burial is reserved for the eminent elders. Local belief has it that common folks, if buried, would contaminate the soil. The wealth of the Maasai is measured in the number of cattle, wives, and children each has.

For special guests visiting them in their homestead for an overnight stopover— *manyatta*—the elder will often, as a courtesy, offer one of his wives who would willingly look to please the guest. The procedure is straightforward: the elder gives his spear to the guest who then surveys the wives and picks one. She would then lead him to her *manyatta,* whereupon the guest will drive the spear into the ground by the entrance, signifying "*do not disturb.*"

History is maintained orally by elders telling stories around a small fire in their *manyatta* where roasted meat and a locally brewed honey drink is passed around. Youngsters are required to repeat and this is how contact is maintained with ancestors. The Maasai live off their cattle; meat is roasted, milk and blood are mixed and drunk. They slaughter cattle as a celebration for special occasions.

The Maasai go through coming-of-age rituals as an essential rite of passage for boys between fifteen and twenty-five years of age when they go through circumcision and are then classified as having advanced from boyhood to junior warriors, the *moran*. Circumcision is performed by elders using traditional knives without anaesthesia.

The candidate is supposed to show courage by not crying or showing pain—else he brings shame on his family. In years past, a boy was supposed to kill a lion to qualify for circumcision, but with strict wildlife protection measures in place now, this practice has been abandoned. They will undergo another ritual from *moran* to junior elder. The females also undergo circumcision, better described as genital mutilation, as a rite of passage from girlhood to young womanhood.

Kalenjin

Lance Corporal Alphonso Kimutei, a Kalenjin, was a gem amongst our men. He would narrate amusing stories at a *ngoma* in the evenings and was the spark of the camp.

The Kalenjin number around five million, representing fifteen percent of Kenya's population. They live around Mount Elgon, Lake Baringo, and the Uasin Gishu district. They are closely related to other smaller tribes including the Nandi, Marakwet, Tugen, Pokot, Keiyo, Kipsigis, Samburu, and Turkana. They are pastoralists by tradition, and slowly getting into farming. Raising cattle is their main occupation and a day's routine starts off with taking cattle to grazing lands and watering holes, and returning to their *manyatta* by dusk. The Kalenjin are a warrior-like people and being cut off from the mainstream and neglected by the central government means they've remained close to their traditional lifestyle.

Their mode of commuting is often by running to attend school which would perhaps be ten kilometres away, and returning to their *manyatta* by dusk.

They had to be in school by sunrise, meaning 0700 hrs, without incurring the wrath of the Christian missionary schoolteachers who would inflict corporal punishment mercilessly if they arrived late. And getting home at day's end during daylight was a matter of survival—hence, the local saying, *"Before the hyenas got to them."*

Turkana

A Turkana, Private Peter Kiplagat, would tell us about his tribe's semi-pastoral lifestyle. They raised camels, goats, and donkeys. Living in arid northern Kenya along Lake Turkana forced them into a life of hardship. This was a consequence of poor, dry weather and the scarcity of grazing land. With few rains, the area is often dry, forcing them to dig shallow, open pit wells in riverbeds for water. Men would always carry a three-legged stool as the sandy ground was too hot to sit on.

At night, this stool was used as a pillow to keep their heads from insects on the ground and maintain the shape of their coiffured hairstyle. A young man would be given a goat to start off his herd. Livestock would equate to currency, and was used to pay the price for a wife. The bigger the herd, the more chances for a Turkana tribesman to have several wives and many children, as polygamy is part of the culture. Turkana rely on their cattle for meat and drink milk mixed with blood as a source of nutrition.

Baluhya

Lance Corporal John Masinde belonged to the Luhya ethnic group, and was very much into himself. He preferred loneliness and to be left alone. This was very much against a team-spirit culture and we had to bring him into the fold as he was a highly self-conscious person who always delivered upon and exceeded expectations, but preferring to do tasks by himself.

He had to be brought into the mainstream of our platoon routine and the only way for me to do that was to assign a buddy of the same rank, Lance Corporal Wambua, a Kamba whose gifts of storytelling and humour became legendary. He was the right person to befriend Masinde, and it did not take long before Masinde came out of his shell and started participating in group discussions, volunteering for patrol assignments, and generally making himself available to lend support.

Other than supporting him, we learned a lot from him. The biggest reward was the lesson that a person can be moulded from isolation to an effective team player with the right guidance and encouragement.

Masinde came from Kitale, in western Kenya. He would entertain us with the Luhya folklore. Marriage amongst the Luhya, including his own, was arranged by the parents visiting the parents of the girl and negotiating a dowry.

A typical requirement for a bride would be the ability to cook, have children, and work the fields growing vegetables. His parents paid a standard dowry of ten cattle and a similar number of goats and sheep. Once the dowry had been paid, his sisters and nieces went to the bride's home and fetched her to begin a new life as a wife. Children are named after ancestors, beginning with the paternal and maternal grandparents.

At home, the man would be the final authority. Cases of larger disputes with neighbours would be taken to a council of elders who would elect a headman. His findings and judgement were final and not subject to negotiations.

Being polygamous, a man would have several wives, but the authority is held by a male and would be passed down from the son of the elder wife. A middle-aged man would have, on average, three wives. Masinde had been married two years and was looking to save enough money to get a second wife. In the Luhya culture, when the old man passes on his authority to his sons, the sons get a young bride for him to marry. After deployment, Masinde's goal along with that of his three brothers was to find one such bride for their aging father. This young bride would most likely be one who had children out of wedlock and had not been married before.

Burial had a special significance for a wealthy person. An old, large tree would be uprooted, and the deceased would be buried there. The event would take on an almost festive mood that was more about celebrating the person's life than mourning his death. To replace the uprooted tree, another tree would be planted by either a virgin or an old woman. The mourning period would last thirty moons: a month.

Inheritance had a well-defined form. A widow would be inherited by the eldest son except if she was his mother, or often by uncles or cousins.

The Luhya worshipped the god of Mount Elgon. A popular Luhya medicine man and healer, Elijah Masinde, started a cult in the 1940s called *Dini ya Msambwa* that soon had a following that continues to this day. About the same time, Christian missionaries came to western Kenya in the early 1900s, and started conversion programs by setting up schools and medical clinics, and conducting church services on Sundays. This had a wide appeal to many, and Christianity spread fast.

Children were named after ancestors or the weather. "*Njala*" means hunger. A child born during famine would be named as follows: A male would have his name starting with a W, as in "Wanjala," and the female starting with an N, as in "Nanjala," thus both having the same root. Males were circumcised at around age fifteen.

Once circumcised, the male would be part of an age group that numbered eight and each age group was for ten years. A man was supposed to live only up to eighty years and not beyond. He was not supposed to go through a second cycle of the age group. Tradition had it that to streamline the age group initiation, an old person reaching eighty would be killed to avoid him having to repeat the cycle of newly circumcised males. With Christian influences and education spreading fast, such traditional beliefs began to change.

The deceased were buried in shallow graves facing east; some clans buried the deceased in a sitting position. If wild animals like hyenas dug up the remains, the skull would be hung in a tree.

The safety of the skull would be entrusted to an elder, often a woman, and if the family migrated to a new location, the skull would be ceremoniously relocated, implying that the spirit of the ancestors was still with them.

The roles of males and females are clearly defined in the Baluhya culture. Whilst a female's duties are to run the household and raise a family, a male's role is to herd cattle, defend the tribe, and hunt for game. A council of elders acted like a court to settle disputes. Punishment was swift and enforced rigorously. In the event of a major dispute and if an accused male was found guilty, his property would be distributed to the family of the aggrieved party as compensation. In serious cases, the accused would be thrown out of the community and forced to go far away from home to live with other clan members and start life afresh. Cattle not only signified wealth and social standing, it represented currency for dowries and sources of barter for land.

Witchcraft

Witchcraft has been a part of culture of many tribes. Its beliefs with superstitions and fears equated to a religion resulting in a devoted following. Politicians, government officials, and business leaders consult the local medicine man, the *maganga,* for advice on major decisions. It has flourished over the years along with Islam and Christianity. Witchcraft is both feared and practised by many tribes. Many would resort to seeking guidance from a witch, who was often a female but could be a male elder, as well.

A well-known medicine man, Juma Tsuma Washe, widely known as *Kajiwe,* hailed from the Rabai district of the coast and exercised much influence across Kenya. Folklore has it that he cured diseases by using herbs and other exotic ingredients to make concoctions, drive away spirits, exorcise demons, and cast spells on undesirables. He was a feared man. His popularity was so strong that, in 1969, the government issued him with a permit to practise his witchcraft. People would go to Kajiwe to be cured of diseases and for cleansing evil spirits. When he died in 1993, none of his children took up witchcraft, but the practice still exists.

Several of my soldiers, regardless of their tribal affiliation, would consult a witchdoctor before taking a major decision. Such was a stranglehold of a witchdoctor on his followers.

MILITARY TRAINING SCHOOL, LANET, KENYA

My recovery from wounds suffered during the Merti Operation was speedy. I was promoted to captain. I was feeling uncomfortable limping with a cane and was constantly inviting sympathy, much to my unease. I was offered a desk job in army headquarters as a staff officer or a training appointment with less physical demands. I opted for the latter. I was posted to the Military Training School (MTS), East Africa, in Lanet in the Rift Valley, to organize, launch, and run an officer cadet training school under a new wing called the cadet training wing. The MTS had been a small but important training base for the KAR that was now going to be developed into a major training centre.

The instructors at MTS were previously drawn from the KAR units in Tanzania, Uganda, and Kenya. We had a shortage of officers and NCOs as the Africanization process of officers and NCOs was proceeding at high speed. We had a fair number of British personnel who had been seconded to the new Kenya Army, and this ensured the continuity of the high standards in administration and a smooth transition. As the armed forces grew, MTS would undergo a massive expansion, changing names to the Armed Forces Training Centre to, as of 2016, the Kenya Military Academy, which offers undergraduate degrees in association with Kenyatta University. This has been a huge development from the humble rudimentary beginnings where text material was typed on Stylex machines and there were few reference books shared since there was no library.

Finding candidates for officer training posed a massive challenge that had to be addressed quickly. We embarked on a country-wide campaign to promote careers in the military to high school students.

The highest academic qualification upon leaving high school was the Cambridge Overseas School leaving certificate. Nairobi University, which was to become Kenyatta University, was itself in an embryonic stage. Students there were more inclined for professional-degree courses leading to law and engineering and seemed to stay away from pursuing a career in the armed forces. As such, we ruled out the university as a recruiting ground.

Hence, our recruitment base was the high school. I went to numerous schools across the country and spoke to students in their last year, highlighting the prestige of a career in the army and its benefits of accommodation, medical care, generous pension at retirement, and opportunity for advancement in numerous skills and trades.

The response was amazing. Students were enthusiastic to follow a career in the army and we got encouraged that the quality of future officers would be high. Until then, the small officer corps had been hurriedly promoted from NCOs whose exposure to formal classroom teaching was limited to whatever education the army had provided.

TRAINING OFFICER CADETS

The new Kenya Army paraded in Nairobi and major towns in attractive red tunics accompanied by an excellent army brass band. They played popular tunes such as "*Tu Funge Safari*" ("pack up your bags to go to war"*)*, stirring emotions and becoming popular with the public. The ongoing Shifta insurgency in the NFD and the regular news reports of successes around it portrayed a macho lifestyle for a soldier. The traditional warrior instincts of students who were largely the first generation from villages to go to proper schools was encouraging, and made for easy recruitment. Applications for recruitment came in large numbers.

This would require a proper evaluation process with selection of candidates done on merit. Wherever possible, we had to aim for a fair ethnic and regional representation. I was appointed deputy to a local and recently promoted Major Titus Wambua, who was the officer in charge of the cadet training wing at the MTS.

We were assisted by two seconded officers from the British Army, Major (later Colonel) Brian O'Hara, an artillery officer, and Major Ian Tomes, an infantry officer from the Warwickshire regiment. We were mentored by an experienced officer, Major Edward Evans of the British Army, who had been at the MTS during the KAR days for many years and was a seasoned veteran of a training routine. The school was under the command of Lieutenant Colonel Abdirahman, whose recent promotion had been fast-tracked to fill this appointment. We had a free hand to develop a curriculum from the remnants of whatever few training guides existed. Prior to the new cadet wing, MTS primarily trained NCOs and personnel from the Kenya Regiment with a training focus on barrack drills, weaponry, tactics at the platoon level, and unit administration.

We were now to train officer cadets who would be the future senior commanders. This meant developing a new curriculum. We had no previous experience or manuals to guide us. The three of us, with help from Edward Evans over many late-night sessions, developed a six-month curriculum. Our own recent experiences and what we remembered from our days at Sandhurst (which some of us had attended) were helpful.

We split the course curriculum into the clearly defined segments of leadership, command and control, regimental organization and unit administration, medical and logistic support, transportation, interaction, and liaison with the police and administration. We did not cover the politics of Kenya or any international affairs, as both subjects were sensitive with accusatory fingers constantly being pointed at the ills of colonialism.

Covering an event in far-off lands or references to battles of WW I, WW II, or the Korean conflict, even to draw lessons from their failures and successes, was of little interest. At this early start of a career in a newly emerged country, the most likely role for the army was local, confined to internal security. Plus, MTS was in its infancy and we had to work with limited resources.

The new army was in an embryonic stage. We had neither a military doctrine to guide us nor external strategic interests to defend. Our role remained to maintain stability within our borders. We were not part of any regional or international military alliance. Our emphasis was to focus on local events closer to home, such as the Shifta insurgency, and to lend support to police services for law enforcement. It extended to handling periodic threats from nomadic tribes in the neighbouring countries where cattle-rustling across the border had been an ongoing menace. Our doctrine remained to be prepared for internal security and counter-insurgency operations. We addressed other likely roles such as assisting in natural disasters, but not much beyond because we had no resources.

We maintained that, as MTS grew and we attained a measure of stability in our organization, we would expand the curriculum to study other campaigns, widen our academic horizons, and become international. We were to be aware of issues and developments in current affairs only from a general-knowledge perspective. Our goal remained to transform cadets from raw civilians into soldiers, with rigorous basic training followed by education on service discipline, regimental values, customs, traditions and the instillation of a regimental pride. Character and leadership qualities had to be developed, and general knowledge had to be widened so that the future officer was well equipped to easily adjust to changing circumstances.

In time, as the army expanded, our curriculum would be revised to address new realities such as joint training with engineers, paratroopers, artillery, armour, air force, and the navy.

We presented an outline of our plan to the army commander, Brigadier John Hardy, who gave us the go-ahead to develop a detailed curriculum. We were overjoyed at this huge responsibility to implement our plan.

To start with, we had to develop a selection process that would see a large number of enthusiastic candidates undergo rigorous evaluation, selection, and training. Our experiences at Mons, Westbury, Sandhurst, and Warminster came in handy here. Combining whatever material was in our possession, our recent experiences in the UK, and assistance from the British High Commission, we developed basic, rudimentary training booklets, manuals that were cyclostyle-copied and stapled. It was a fantastic opportunity to plan and run a key training institution from the very beginning, with limited access to resources, to prepare future leaders. We tapped into an abundance of improvisation and creativity.

We were supported by an enthusiastic team of dedicated WOs and NCOs who were the backbone of our efforts. As candidates went through a rigorous, demanding basic-training regime, we developed an evaluation process with independent input from all the instructors.

We would meet every Friday morning to grade each cadet. Such evaluation was based upon a wide criteria of knowledge, participation, team effort, initiative, leadership, enthusiasm, and soldierly skills of drills, weaponry, and general knowledge of current affairs in Kenya and its neighbouring countries. We kept away from any reference to, or discussion about, politics or ethnicity for concerns about inviting divisiveness.

Happily, after six months, our first batch of fifteen new and locally trained cadets would be commissioned at a dignified parade with the first locally appointed commander of the new Kenya Army, Brigadier (later Major General and chief of defence staff) Joseph Ndolo, taking the salute.

Accompanying Army Commander Brigadier Joseph Ndolo at the first passing out parade of locally trained and newly commissioned officers, Lanet, Kenya, 1965.

We organized the commissioning parade, normally referred to in military jargon as passing-out parade along the lines of the Sovereign's Parade at Sandhurst, with a superb Kenya Army brass band playing *Tu Funge Safari* and other regimental marches to much applause from invited guests and spectators. After the parade, we retired to the officers' mess for a well-attended reception. It was wonderful, and most satisfying to see that from this first batch of locally trained officers, a few rose to senior appointments. In later years, some held the rank of brigadier and some rose to lieutenant general, appointed to command the army and another to command the young navy. It has been most gratifying to see our efforts with meagre resources achieve such remarkable results. The MTS has grown over the years from humble beginnings to the present-day Kenya Military Academy, offering a wide range of courses up to an undergraduate degree.

DEFENCE HEADQUARTERS, STRATEGIES AND PLANNING

From MTS, I was posted to defence headquarters in Nairobi in 1966, as a staff officer in the operations and training branch, my speciality—or so I thought. This was a key appointment since the ongoing Shifta insurgency had to be tackled with a whole-of-government approach. The defence headquarters' organization was in its infancy, and various branches were being formed to provide an efficient administration and management of the armed forces. I worked directly under the principal staff officer, Colonel Lucas Matu, who was technically the chief of staff, and was advised and guided by the seconded British officer, Brigadier John Anderson.

As a senior operations staff officer, accompanying Army Commander Brigadier Joseph Ndolo and chief of staff, Colonel Lucas Matu on a battalion group exercise in Nanyuki, Kenya 1966.

I was responsible for liaising with the diplomatic missions who were helping us and coordinating the annual inspections of battalions and support units who were tested on their operational preparedness. The assignment exposed me to working with decision-makers right at the top level. From the military side, I represented the armed forces on the National Security Committee. We would meet once a week to assess intelligence reports and recommend a response with a plan.

RISING FROM THE DEAD

In the mid 1960s, I had my sights on attending a mid-career course overseas. We received many offers from Australia, Canada, the United States, and Britain. My preference was Britain. In preparation for an opportunity arising, I wanted to regularize and have my travel documents prepared. Until then, I was, as were most Kenyans, still a British citizen who had not yet taken out Kenyan citizenship. I had filled my application forms for Kenyan citizenship upon my return from the UK in early 1965, but had neither heard back from the authorities nor followed up. On a casual encounter with the principal immigration officer, Bernard Bangua, at a function at our Waterworks camp in Nairobi sometime in 1966, I mentioned my application for citizenship. I said that I had not heard back and that I needed to apply for a Kenyan passport. Bernard was in charge of citizenship issues and it was his department that vetted applications to ensure that they met the requirements for such a certificate.

Bernard called me to his office at Jogoo House in downtown Nairobi a few days later and said that something was amiss with my application. He told me that there was no record of it in the pipeline. But on checking further, he said that one bearing my name was killed in action in Merti on April 20, 1965, and that my application for citizenship was terminated. I could not believe what I was hearing. He was equally suspicious that something was not right. Had it been a case of paperwork getting lost in a young bureaucracy?

He asked me to accompany him to the basement. After a relentless search to locate my file, I was shocked to see a file folder with my name on it tied up in a red band and marked "DECEASED" in large capital letters. Bernard, just as much in a state of disbelief as I was, jokingly broke the silence with: "You have risen. . . . Resurrection, my friend!" We both laughed. He undertook to regularize my paperwork and assured me that everything would be fine. He found out that when I had been wounded in action in Merti, as the battle was going on, the first signal sent by my wireless operator, Private Ngugi, to company headquarters that had miraculously reached army headquarters had mentioned that I was "KIA," or *killed in action*. In the operations room where Slidex and Morse messages were received and directed to various departments for action, mine had been sent to the section that handled burials. Somehow, my message had found its way to the immigration department and, hence, the mystery of my short-lived fate.

I could not believe what I was hearing because I was still receiving my salary and was very much alive. Indeed, I had only recently been recommended for a Distinguished Service Medal for my combat action, which was to be awarded later that year at the State House where Jomo Kenyatta had conducted the first state investiture in October 1967. Upon seeing me amongst the recipients when we were rehearsing for the parade, he walked over and said, "*Kijana, wewe ta linda,*" or "Young man, you will lead." It was humbling remark from the head of state and commander in chief.

My citation read:

"On 20th April, 1965, a platoon (1st Battalion, Kenya Rifles) commanded by Captain Sethi was pursuing a gang of thirty well-armed Shifta in the Merti area. During the action, he was wounded in the thigh by a rifle bullet, and shot his assailant dead. Then, despite much loss of blood, Captain Sethi continued to direct operations until he fell exhausted. He showed exemplary devotion to duty in this action—his first operation against the enemy—in which five Shiftas were killed and five others mortally wounded."

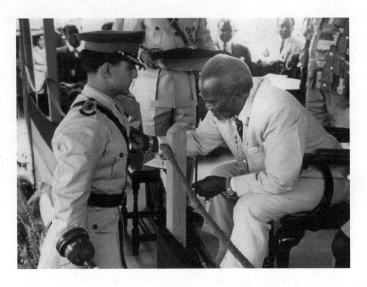

President Jomo Kenyatta pinning the Distinguished Service Medal, at State House, Nairobi, 1967.

But the situation had to be rectified with bureaucratic paperwork and this meant that I had to have a notarized document confirming who I was, witnessed by someone who knew my family and a little about my present and recent past. Bernard undertook to have my papers regularized at the ministry of defence. Sure enough, after a few days, I received my new Kenya citizenship certificate number 9248 dated May 24, 1967. I did not find out if it was one with the same original serial number or if a second one had been issued. I was just relieved that I could now apply for a Kenyan passport. I looked at the entire episode with a mixture of uneasiness and nervousness. In its course, I learned the fate of an individual who dies on active duty, with the files safely locked in a basement somewhere for dust to collect and their contents to be forgotten. Such is the unpredictability of life and the even greater mystery of what lies in store when life comes to an end, which it must for us all.

AID TO CIVIL POWER, NYANZA FLOODS, 1967

I soon had my first practical taste of *"aid to civil power,"* a subject to which we had been introduced at Sandhurst. But other than a classroom academic exercise, I had no field exposure on this. Our efforts remained focussed on fighting the Shifta and we kept a close watch through intelligence services on any security-related developments in the rest of the country. Our priorities lay in building a well-rounded armed force from the nucleus of the King's African Rifles. As such, we overlooked checking whether there were any contingency plans or resources to cope with a natural disaster where the army could be called in to assist.

There were no such plans. At MTS, we addressed the subject of aid to civil power as an academic exercise in the belief that it would remain as such and not knowing that we would be very much involved.

There was no provision for strategically stored emergency stockpiles or the formation of committees from security and administrative services to ensure command and control, lines of communication, and some kind of liaison among the numerous aid agencies. There was no system of early warning of potential disasters. The meteorological service was in its infancy and, as such, the forecasting of weather patterns of heavy rains and drought was not developed. A cyclical pattern of short and long rains with near-accurate dates was the closest to forecasting weather patterns that we had, but this did not account for their severity.

Aid agencies such as the Red Cross operated in consultation with the ministry of health and the church. Otherwise, they enjoyed a free hand to provide few basic health-care services in rural areas to complement those provided by rudimentary health clinics. There was negligible interaction between the aid agencies and the security forces.

In early 1967, heavy rains had caused havoc in the western province of Nyanza. Rivers Nyando, Nzoia, Awach, and Sondu burst their banks, destroying infrastructure, displacing thousands of people, and resulting in several deaths. The hardest-hit areas were in the low basin, which thrived in agriculture since the soil was rich. Here, villagers had their rudimentary dwellings with cattle in their farmsteads. The rivers in close vicinity were the source of water for drinking and irrigation. The main crop was sugar cane, with the two refineries in Muhoroni and Mumias processing cane and refining it into sugar for domestic consumption and export. The sugar refineries came to a standstill, since workers could not get to work and farmers could not transport cane to the refineries. Granaries were destroyed and grain stocks swept away. Bridges were washed away, too, and mudslides closed roads, cutting off communication and leaving communities isolated. Telephone services were also cut off. A lucky few escaped to higher ground, but were marooned and helpless.

Boreholes—open, shallow latrines—were either submerged or washed away, posing health risks on a huge scale. Water was contaminated and fears of a cholera epidemic and diarrhea abounded. There were fears that malaria and mosquito-borne diseases would spread with catastrophic consequences. An invasion of the locusts who thrived in such conditions would make matters worse. Livestock was swept away, posing further risk for disease. This was a tragedy of immense proportions. Since the water table by Lake Victoria is shallow, flooding and washed-away open-pit latrines and shallow wells caused rapid contamination of groundwater.

Added to this was internal ethnic fighting that had gone on for generations during floods, famine, and drought. A worsening security situation that would adversely affect access to basic services loomed. Some villagers took their cattle to high ground. Traditional hut dwellings were made of poor materials and mud; they could not withstand marshy water conditions so simply got swept away.

Small rural clinics in villages and towns that provided basic medical-support services ran out of supplies, and were not able to cope with the large number of sick who struggled to reach for help. With a lack of food and inadequate health-care support, it was only a matter of time that the catastrophe would lead to famine. Such adverse economic consequences would further produce social upheaval, widespread disease, and political instability.

Sadly, with no contingency plan to handle such a situation, the army was called in to assist. The chief of staff, Brigadier Lucas Matu, hurriedly composed a team of staff officers and field commanders to plan a support program to assist under an "aid to civil power" banner. There were no manuals to guide us, and we had no experience handling such an operation on a national scale. We did what little we could out of sheer initiative and creativity. I was part of a small team organized to plan and execute a support mission; in two days, we had the situation under control and troops were on the way.

We sent paramedics equipped with large stocks of quinine and antidiarrhea pills, a large fleet of cargo and water trucks, water-purification tablets, food supplies, blankets, and mosquito nets. We also set up several evacuation and support centres.

The Kenya Air Force, assisted by a well-organized Police AirWing, would drop food supplies on higher ground. They also provided intelligence from the air, which assisted us in directing aid.

I was the field liaison officer charged with evaluating and coordinating the situation in consultation with the provincial and district administration officials, police detachments, the Red Cross, and a few volunteer organizations who had arrived to assist. The response from first-line responders was amazing.

We set up a command centre on the grounds of the Mumias Sugar refinery and began to work. First, we identified several safe centres on high ground where displaced people had gathered. We organized support teams consisting of a medic and two others and headed by an NCO with the task of determining the type and extent of assistance required. We carried a large stockpile of tents, blankets, mosquito nets, water-purification tablets, canned food, basic medicines, and even cooking utensils. We set up evacuation centres at several safe places and provided tents, blankets, and food. We sent several water trucks and positioned them on high ground to provide safe drinking water.

The administration officials, local community leaders, chiefs, and headmen were invaluable in their assistance and we made sure that they remained as the face of government as we worked with them. In the eyes of the people, they had to be relevant and seen to be in charge. For political reasons, we had been instructed that our mission was to provide support without coming into the limelight.

The government sent ministers and officials to address those affected with assurances of support. At a large gathering, I met the justice minister, Tom Mboya, who was most gracious in his acknowledgement of the army's support.

As with any politician looking ahead to the next election, he wanted to be seen by his constituents as not having ignored their plight.

I had met Mboya briefly around 1962 when I had applied for a scholarship to study engineering in the United States and got accepted, but later declined when I joined the KAR.

We handled the agricultural calamity as best as we could. Where conditions allowed, workers manually cut cane with slashers and piled it on the roadside so army trucks could take them to the refineries. The object was to salvage agricultural produce wherever possible and get employees back to work. The goal was to restore electricity, commercial activities, and routine as quickly and as best as we could.

Within about two weeks as rains subsided, the situation got better. The government took note to be prepared for such an eventuality in the future. It set about to create contingency plans to properly handle such natural catastrophes right across the country. Disaster-relief committees were organized up to the district level. Their task would be to equip rural medical clinics with adequate emergency supplies. For municipal-infrastructure improvements, recommendations were made to construct sewage-treatment plants and piped drinking water.

Priority was given to improve the de-siltation of river banks and streams to allow a smooth flow of water to Lake Victoria. Villagers were encouraged to construct their new houses with firm structures on higher ground and consider new crops in low-lying areas such as banana and maize, which could withstand future torrential rains.

When the floods subsided, a thorough evaluation of the government's response was carried out to ensure that a policy was in place to cope with such calamities in the future. A disaster preparedness and management response unit would be formed with first-line responders and a well-defined command-and-control mechanism.

BRIEFING KENYATTA

Defence headquarters exposed me to planning an expansion of the armed forces and coordinating financial, operational, and procurement issues. Such challenges were strategic at a national level and had to be addressed promptly. The new Kenya Army was formed from the remnants of the KAR, which was undergoing a major overhaul from every aspect. It was primarily a basic infantry force, and it needed complementary combat support and logistic units.

The treasury had already informed us that there was no funding for equipment procurement and, as in any bureaucracy, acquiring funding would be a lengthy process. We had to submit a forecast of expenditures that would be scrutinized and, if treasury officials were satisfied, included in the annual national budget. The budget would be debated in parliament and, should our estimates pass, we could then go through a tendering process. The timeline for this exercise was two to three years to secure funding; another one to two years for selecting bidders and placing orders; and, depending on the type of equipment, armaments, ships, or aircraft in question, another several years including training on the equipment before it could be brought into service. We had then to plan and conduct joint exercises to ensure operational preparedness. Various units with new equipment and their own logistical support had to be extensively trained as a cohesive body with the command-and-control mechanism organized properly to efficiently mount operations.

This was a challenge of immense proportions that many of us who had risen rather fast in our careers were not experienced enough to handle. The chief of staff, Brigadier John Anderson, came up with a brilliant idea: we would bypass bureaucracy. We would make a detailed presentation of the security threats we were facing as a national priority to the president, Jomo Kenyatta, who was also the commander in chief of the armed forces. We would clearly define a defensive capability we needed to counter such threats, highlighting the worsening situation in the NFD. We would make this presentation with the support of charts and timelines as a team, and would include the head of intelligence, James Kanyotu, and the chief of police, Bernard Hinga.

The presentation took place at State House, Kenyatta's official residence which doubled as office of the President and residence on the outskirts of Nairobi. Our presentation team was headed by the chief of defence staff, Major General Joe Ndolo; the chief of staff, Brigadier John Anderson; the principal staff officer, Brigadier Lucas Matu; myself; and four others. We were selected to prepare various five-minute presentations.

My task was to highlight the lack of capability we were facing from an operational perspective; basically how ill-equipped we were regarding equipment needed to mount operations for the army, air force, and navy.

It would be from here that a doctrine would evolve that would ensure that the armed forces had the tools for law enforcement to combat the Shifta insurgency, participate in any internal security operations to secure the region, and foster stability.

It was a tall order indeed for a twenty-four-year-old captain, launched into the dizzy spheres of high-level policy development with no preparation or experience.

Just to have an opportunity to address Jomo Kenyatta and top-level cabinet ministers and select bureaucrats in a closed, confined setting of around twenty was more exciting than the presentation. This was also a once-in-a-lifetime chance to shine before the commander in chief, and such thoughts were quite normal for career advancement. A thorough preparedness was necessary, since it would mean either going up the career ladder or falling by the wayside. I prepared my briefing notes and showed them to John Anderson, who made some changes. I practised the revised version several times to make sure my delivery was professional, convincing, and articulate. Five minutes seemed a long five minutes.

The presentation moment arrived. Ndolo made an introduction, followed by Anderson, who described the broader picture of a worsening security situation, followed by James Kanyotu, who said that his intelligence showed that the Shifta insurgency was gaining momentum and that neighbouring Somalia had increased its support to the dissidents. I was the fourth speaker, and I highlighted the need to mount operations aggressively to contain the Shifta insurgency.

We lacked reliable field radios, reinforced troop-carrying vehicles, armoured cars for recce, mine-detection equipment, and improved casualty-evacuation means, including aircraft. These shortcomings, along with several others, had to be addressed as a priority.

Jomo Kenyatta was a larger-than-life person. He had an inspiring personality mirroring a majestic global image that he had acquired through his struggle for independence. Most importantly, his views to build a new multiracial society without bitterness about the injustices of colonialism had earned him respect and donor support with soft loans, grants, and donations. Kenyatta had a magnetic appeal and carried the authority of not only a tribal elder, but a world statesman. His deep, penetrating eyes, and charming, joyful personality radiated a hypnotic, magnetic aura that was reflective of his images in newspapers. Just being in his presence was special, and addressing a select gathering with him on centre stage was a privilege of immense proportions. After I spoke, he gracefully acknowledged me by simply uttering, *"Mzuri sana, kijana,"* or "Very good, young man." Our thirty-minute presentation was over; we packed our charts and left. On the way out, we curiously checked with each other about how we had fared.

The team must have done a good job, since only a few days later, we got instructions to prepare a shopping list with specifications of what we needed. We sent requests for support to several potential donors and, much to our delight, their response was positive. A lot of government-to-government liaison must have followed. Various countries rushed forward with attractive offers of donations and generous loans to supply us with equipment. The aging fleet of British Bedford Vauxhall trucks was replaced by German Daimler Benz trucks, the Marconi and Racal wireless equipment was replaced by French Thompson CSF, Britain donated a fleet of Ferret and Saladin armoured scout cars and 125 mm towed artillery guns, and the standard British soldier's 7.62 mm Enfield rifle was replaced by German Heckle and Koch 5.56 mm assault rifles.

An engineer unit was formed with such basic skeletal equipment as heavy-duty Caterpillar graders, Bailey Bridge kits, and assault boats for operations in lakes and rivers.

Slowly, these units grew to battalion strength. We introduced a parachute unit for political expediency. The young air force received the Canadian De Havilland Buffalo aircraft to complement its fleet of six Beaver and four Caribou aircraft. These were later used to carry cargo and transport troops. We expanded aircraft maintenance facilities at the base in Eastleigh and similar naval maintenance facilities at the naval base in Mtongwe, Mombasa.

The young Kenya navy, with just three small aging patrol boats, got a boost. With an inconsequential defence capability but tasked with enormous coastal surveillance and law-enforcement roles, the fleet was upgraded with orders for four 120-foot patrol boats made by Vosper Thornycroft of the UK.

Such a mammoth expansion required enlisting suitable manpower. These individuals would be trained locally and abroad in the use of specialized equipment, giving our skeletal maintenance facilities a boost for a well-equipped domestic-service capability. Additionally, we thought from a strategic view that, if we could build state-of-the-art workshops to service specialized equipment for the army, air force, and navy, we could become a service centre in the region to provide such services to our neighbours. The British High Commission maintained a defence advisor to liaise with us and, subsequently, France, Germany, the United States, Israel, and India rushed in to appoint similar defence attachés in their embassies. Looking back, it is natural to have pride for a task of such immense proportions entrusted to a few of us who handled it successfully with professionalism and utmost enthusiasm, along with much support from donor countries.

DESTINY BY ACCIDENT—
AUTO DEALERSHIP

The expansion program of our armed forces was now underway. During the changeover of our British Bedford Vauxhall troop and cargo-carrying fleet to German Mercedes Benz trucks, I met and worked closely with the Daimler-Benz dealer, David Dobie, under whose name the dealership, D.T. Dobie and Company (David Theodore Dobie), operated in Nairobi. Aside from our contractual negotiations on a government-to-government basis with Germany handled by our ministry of finance, we liaised at the local level to implement the contract with Daimler-Benz. A wide range of activities had to be addressed, including training, stocking parts, and upgrading our workshops with specialist tools. I got to know David quite well. In the beginning of our relationship, he was discreet and rather reserved. In time, however, he accepted me as a colleague he had to work with. Our relationship became somewhat social.

I was fascinated by his breathtaking stories of WW II and the daring role he had played. At first, he was reluctant to relive them, as he always felt guilty about the many of his colleagues who were killed in action and not fortunate enough to return to their families after the war. He felt a responsibility for some he had to abandon on the battlefields dying, gasping for breath, and desperately seeking help with outstretched arms and deep, penetrating eyes. Such reflections always brought sadness to him and he would simply say, "I can't go on. . . . Let's change the subject." The more I saw him, the more I got closer to him because there was something special about him that he kept to himself. His wartime stories were a mixture of suspense, bravery, camaraderie, loyalty, leadership, and, most importantly, acknowledgement of the vulnerability of a human. As our relationship deepened, he would narrate tales of his extraordinary life and how he ended up in Kenya running a successful business.

I started looking at him as a role model, convinced that if ever I chose to leave the army, I could also follow in his footsteps to any other profession in any country and be as successful.

What was important was to remain focussed, to display honesty and integrity, and to build a sound reputation. Earlier teachings came to the fore: never to forget that, aside from the products to sell or the services to provide, the most important factor mirroring you is the respect you give and earn with people you work with, being a team player, and accepting that people around you are both weaker and stronger than you, but always part of a link, much like in a chain. The human factor is the most important.

David would relive his experiences, often over scotch and soda in the comfort of his palatial home, in bits and pieces. His life during WW II was full of action. He commanded a parachute battalion of the British Army in 1944 during its advance to Arnhem in the Netherlands as part of Operation Market Garden, where they had to capture the crossings over the River Rhine. His unit suffered heavy casualties with the Nazis in heavy pursuit. He was wounded, and took shelter in a remote farmhouse where he was taken prisoner. While being treated for wounds, he made a daring escape through enemy lines and eventually rejoined his unit. On one incident, when he discovered that some men from his division were trapped and hiding behind enemy lines, with radio silence enforced, he swam across the frigid waters of the Rhine to alert the British who were on the other side of the river. They were able to mount a rescue immediately using canvas boats and with very little preparation.

David would recall that, when he emerged on the other side of the river soaking wet and dripping, he was initially suspected of being a spy and was nearly shot!

After the war when soldiers were discharged, they were offered several opportunities in the colonies as part of a settlement program. David, in his mid-thirties, had decided to come to Kenya. He was penniless, but this did not deter him from trying different options.

He was not cut out for the farming that many of his colleagues went for.

He was also not interested in civil service or any of jobs available in agencies such as the railways, ports, and so on. He took employment in a garage as a manager and, with his liking for cars, found the job satisfying.

About 1948, by which point he was enjoying his new life and slowly settling down to a satisfying routine, he responded to an advertisement in the local papers inviting candidates to apply for a Daimler-Benz dealership.

At first, he was hesitant since, until recently, he had been fighting Nazi Germany and could not accept that he would now be drawing a paycheque from an enemy who, for him, was still an enemy. He hesitated, but applied on the reasoning that Daimler-Benz had a good reputation for vehicles worldwide and, with the war over, countries would start building relationships with Germany and if he did not try, someone else would take the dealership. The interview was held in a guest cottage at the Norfolk Hotel in central Nairobi. He was probably the fifth in a field of over twenty who had come from all over Kenya for the interview.

The interview board had three German officials from Daimler-Benz and candidates were interviewed for ten to fifteen minutes each. When David's turn came, he walked in, took a seat facing the three, and the interview process started. The range of questions was standard: why do you think you are most qualified to represent us, what business background do you have, what are your financial resources, how much time and money will you commit to the dealership, are you an existing dealer hoping to expand, do you have a technical background in the auto industry, do you have financial resources to set up sub-dealerships in smaller towns, etc. David's response to all the questions was lukewarm.

He spoke in a soft, convincing voice and simply said that he did not have any financial resources and had a limited background in the auto industry. Having been recently discharged from the army after a successful deployment in Western Europe during the war, he took employment in the insurance business in London which he did not like, and had arrived in Kenya with the hope of rebuilding himself.

He said that he had lost a lot of his buddies in Europe and wanted to get far away to another peaceful country. David had an eloquent, convincing way of describing situations, recalling details meticulously. And with his command experience as a lieutenant colonel, he knew how to remain focussed, in control, and not to ramble on. One of the Daimler officials asked him about his army past, and David narrated his exploits without worrying that he may offend them. He spoke about the parachute battalion he'd commanded heading for Arnhem to capture river crossings; he spoke about how, with radio silence strictly enforced, he'd swum across the Rhine to alert the British troops on the other side about the troops who were trapped behind enemy lines and had to be rescued; he spoke about how he was taken a prisoner and how he escaped.

The interview, now probably at the thirty-minute mark, morphed into a discussion about the war. David was now in control with his captivating wartime stories and, most unlike an interview, the interviewers simply listened to him.

The Daimler official asked what the highlight of his wartime experience had been. David responded that his swim across the Rhine to alert fellow British troops to rescue trapped paratroopers who were behind enemy lines stood foremost. He qualified: as a soldier, you rely on your buddy and fight for him as you would expect him to do likewise; there is a deep sense of camaraderie among soldiers, regardless of whether you are pursuing the enemy or vice versa.

The second one was when his battalion was advancing in pursuit on foot over farmlands toward Arnhem, coming across Nazis who had dug in and were bravely defending their position. In one trench, a young soldier in his late teens or early twenties dropped his rifle, raised his hands, and said, "Sir, please don't shoot; Sir, please don't shoot." Despite a tense situation all around, David disarmed him, and took him prisoner, marching him off with other captured soldiers to a camp where he would undergo an interrogation.

The army believed that POWs would provide a rich source of information that would assist in planning operations if properly interrogated.

David continued: "The eyes of the young kid locked in with mine. I saw innocence in him and, for some unknown reason in the thick of battle, with bullets flying all over, artillery shells exploding, and aircraft on relentless bombing runs, I did not look at this kid as an enemy. He was just another innocent soul who was sent to the front lines only as a number, and probably against his wishes." He ended: "To this day, I have never understood what led to my decision when it would have been easier just to have killed him." Silence followed. David and his interviewers just stared at each other for what seemed like an eternity.

The silence was broken by this one official who called in the secretary, and asked her to cancel the remaining interviews and thank the ones who had already attended, saying that the dealership had been awarded.

David was told that the young soldier whom he did not shoot, Gutner, was now facing him with the tables turned. He remembered the incident very well. David did not mention Gutner's last name. As he was closely connected to the Daimler family, Gutner took it upon himself to support David to the fullest, providing technical assistance and financial support, including helping him acquire real estate and build a suitable garage with showroom facilities. David's luck soared and, over time, he expanded the dealership to Mombasa, Nakuru, and Tanzania. The incident confirms a widely held belief that one's actions follow much like a shadow and, in many strange ways, influence how *destiny is shaped.*

As success multiplied severalfold, David embarked on a mission to assist the disadvantaged in Kenya by providing financial assistance. He would channel his support via the church. David had an extraordinary way of interacting with people of all backgrounds and putting them at ease. He rarely bragged about his achievements, and would always put others' concerns before his. This was Colonel David Theodore Dobie, an officer and a gentleman throughout until he passed away prematurely in 1971. I was to hear later that his exploits as a paratrooper during WW II around Arnhem in the Netherlands during Operation Market Garden were included in the TV miniseries, Band of Brothers, and the blockbuster movie, A Bridge Too Far. I was privileged to have met David and drew inspiration from him for my successes in life.

NEW KENYA BATTALION, 7TH KENYA RIFLES

In 1967, a new infantry battalion, the 7th Kenya Rifles, was formed and given wide publicity at its launch at Langata Barracks on the outskirts of Nairobi. The battalion had trained extensively as a fighting unit and was due to go to the NFD on a six-month operational deployment soon after receiving its colours. I was the adjutant of the battalion, and responsible for making detailed plans for the day. Given the significance of the event, timing had to be flawless. Cabinet ministers, senior governmental bureaucrats, heads of diplomatic missions, and some key business men and industrialists had been invited. We were in the full glare of the media that day, and the weather was sunny and picture-perfect. The main gate had been kept open for the public to attend and the spectators overflowed. Our security had been beefed up with support from other units and the special branch.

Kanwal Sethi

Adjutant, 7th Battalion, Kenya Rifles, Langata Barracks, 1967-1968

At the main gate of Langata Barracks, President Jomo Kenyatta was formally welcomed by the commanding officer of the battalion, Lieutenant Colonel James Lengees, accompanied by the chief of defence staff, Major General Joe Ndolo. He inspected a guard of honour to the accompaniment of a bugler playing The Last Post. Colonel Lengees accompanied him, standing in an open Land Rover en route to the parade ground, followed by other dignitaries. At the parade ground, he was welcomed by the army commander, Major General Jackson Mulinge, and he inspected the parade with the army brass band playing stirring music, with *"Tu Funge Safari"* arousing everyone into collective singing. After a presentation of the regimental and presidential colours, guests were invited to the officers' mess for a reception. This was a colourful affair that was given a national significance of pride since it was the first major army unit raised in independent Kenya. I was privileged to have played a part in this historic event.

A few days later, as the duty officer, I faced an unexpected situation for which I was not prepared. I went around the huge 1,000-acre camp in a Land Rover at night with the camp sergeant major on duty, WO Paul Kimani, to conduct an independent surprise inspection of the camp on any illegal activity or intrusion into the camp. Such impromptu inspections were normal. We drove around to remote areas within the camp, as this was a common practice.

One evening, in mild drizzle, whilst driving on the narrow dirt road toward the sports ground at the far end, we saw a pride of lions sitting comfortably in the middle of the road, some 100 metres ahead. They were oblivious to our presence and we were scared. We did not carry any firearms or a radio, as all communication within the camp was by local landline telephone from various access points. We did not feel the need to carry a wireless set on our persons. My fear was that they might charge us. We were in an open Land Rover with just the frame and a canvas top, and would be an easy prey with no option for a safe escape. There was no way for us to keep going as they were not going to move. Going around them was certainly not an option as, by doing so, we would get closer and appear threatening.

Also, the wide and deep storm water drain either side of the road did not give us enough room for safety. Remaining where we were was getting risky, just in case they charged.

The road was narrow, barely wide enough to allow an oncoming vehicle to pass. Plus, the risk of getting stuck in wet, soggy ground added to our fears. It was important to remain calm and communicate with each other in soft, whispery tones. Reversing a Land Rover in the dark of night on a curvy dirt road without benefit of present-day cameras requires exceptional skills and lots of courage.

Lions right in front of us posed a threat with devastating consequences if we disturbed them or they attacked us. Getting out to assist the driver, Private Kibwego, as he might want to reverse was not a wise thing, and we did not consider this option, either.

In a situation like this, it is foolhardy to insist on exercising authority just because an officer is superior to an NCO. But Sergeant Kimani, having had a much greater experience in bush craft and wild game, knew how to extricate us from this precarious situation. He took over the leadership role, whispered to the driver to start reversing very slowly without making excessive engine noise; he would face the back, using a flashlight and alert him if we were likely to slip into the storm water drain.

After a suspense-filled twenty long minutes, we got back to the main gate guardhouse. A warning was issued by phone to places and offices that were manned—the guard house, medical centre, kitchen, communications centre, and canteen—that nobody was allowed to walk in the camp. Any emergencies had to be attended to by a hard-covered Land Rover or an ambulance.

The following morning, I went back to the same place with the same driver, Sergeant Kimani, and two escorts armed with rifles, only to find that the lions had gone. On closer examination of the area, we noticed that they had gained entry via a cut-out opening in the fence. Much to our relief, we never discovered if any baits such as raw meat or traps had been laid by poachers. We became concerned that perhaps we had poachers amongst our troops, and alerted the special branch to investigate. Word must have gotten around. After exhaustive follow-ups, no one was implicated and, much to our comfort, we never had our chain-link fence disturbed again.

ASSISTING IN ANTI-POACHING

As soldiers, we were never far from participating in unexpected roles. We were requested by the Kenya National Parks to assist in anti-poaching efforts, as the problem was getting worse. We had no mandate for such a task. We informed defence headquarters and, since the problem was of a national concern, offered some initiatives providing they did not get us deeply involved since we had neither the manpower nor the resources for such a task. I would liaise with parks authorities initially with the neighbouring Nairobi National Park, since park headquarters were located there, and later at other game parks where we would conduct field exercises as a show of force to poachers.

Conducting such exercises in game parks had its own challenges. On one occasion, we conducted a week-long company exercise in Tsavo National Park. On the last day, we set our camp mid-morning near a marshy place by Mtito Andei, a small town adjoining the park. We left a small detachment to guard the camp. The exercise required an assault to be mounted on an enemy dug-in location on a hilltop some five kilometres away. The enemy had a good observation over a wide area from the vantage of their hilltop location.

On a field exercise with Army Commander, Brigadier Ndolo, left, and chief of staff, Colonel Lucas Matu.

We covered the distance on foot, using cover of bushes and little hills, taking different routes to get to our assembly location. From there, we were to mount a flanking assault. The object was to avoid being seen by the enemy and surprise him. The exercise was observed by umpires from defence headquarters. After a successful assault for which we were rated "outstanding and operational ready," we walked back to base, arriving in the early evening.

The cook was preparing hot soup for the *askari* in a half-cut-out forty-gallon drum that we customarily used as a cooking pot. This was under an acacia tree. I volunteered to assist by stirring the soup with a large wooden spoon, which was much like a small paddle. In keeping with our routine, I would stand by the cook, complimenting the *askari* for their efforts as he cheerfully served them. I would be the last to have my soup with the cook. Whilst stirring, I felt a rubbery kind of a pipe and lifted it out. It was a snake. It had fallen from the tree. The soup was discarded. An uneasiness dominated from this point forward, as most of us had a fear of snakes. Every small noise sent shivers. Walking around to inspect on sentries, I noticed that nobody wanted to sleep, and everyone was alert and fearful of a snake coming their way. I was just as scared, but had to mask my fear to keep a brave face and maintain courage.

The following morning, we discovered that our campsite was about 100 metres from a massive snake pit. I had never seen a snake pit before. The pit resembled a spaghetti-like scene with slithery snakes crawling all over each other. But the *askari* showed their brave soldierly mantle. After nervously but cautiously dismantling the camp, they hurriedly loaded supplies in our troop-carrying trucks. We took a hasty return to Langata Barracks as per our original plan. For an infantryman operating in remote areas, the enemy is also the wild game and all sorts of reptiles, scorpions, and insects. For these, there is really no preparation other than being watchful and observant.

Officers of the 7th Battalion and others based at Langata Barracks were made honorary game wardens of the Nairobi National Park by the well-known and internationally respected director of the Kenya National Parks, Dr. Perez Olindo.

This was a reciprocal courtesy he extended to us as we had made him and a colleague of his, Julius Mulandi, a game warden, honorary members of our officers' mess.

This was meant to build good relations since the dividing chain-link fence between our camp and the park did not live up to the saying, "good fences make good neighbours."

We and the parks security would often find the chain-link fence cut and could only assume that some mild poaching from our side by some of our soldiers may be responsible, as outsiders entering the barracks had to check in prior to entry.

If there was poaching at all, it was confined to small game such as rabbits and the odd gazelle, but not big game. Lions roamed freely in the game park, along with rhinos, giraffes, zebras, and wildebeest. Finding those lions in the camp was an isolated incident.

As honorary game wardens, not much was expected of us other than to enjoy free entry privileges with a guide and radio-equipped vehicle that belonged to the game park. On casual encounters, Perez Olindo would mention the odd case of poaching at other game parks, since poaching until about the early 1960s was gaining a foothold but had not developed to menacing levels. Poaching still remained a local nuisance, but there were indications that it was only a matter of time before it would become a national threat to tourism. However, demand for elephant tusks, rhino horns, game skins, and animal body parts for a variety of medicinal or health concoctions increased in the late 1960s, especially from the Orient. This resulted in widespread poaching, with poachers using rudimentary traps and bows and poison-tipped arrows to kill wild animals to meet an ever-increasing demand.

Seeing huge profits with minimal risks, the poachers started arming themselves and posed greater threats. The situation slowly became catastrophic, and many international agencies got involved to raise awareness and strengthen laws to protect wildlife and raise funds to fight poaching. The army and the police assisted in law enforcement by sending armed patrols to ward off poachers, but this was not enough.

The poachers would simply wait out until the patrols were gone. A need arose for a well-armed force to combat poaching, and this resulted in a new unit, the Kenya Wildlife Services, or KWS.

This was headed by the world-famous environmentalist, Dr. Richard Leakey. We would often send armed soldiers to accompany the game wardens if they suspected that poachers had entered the park. Such was the dual role of a soldier.

AGRICULTURE SOCIETY SHOWS

The Agriculture Society of Kenya, ASK, representing the farming community, organized annual shows much like county fairs in major towns. There, farmers would showcase their animals and produce. The event resembled the Oscars, and was popular, always enjoying huge attendance. Local manufacturers of various products would also participate and promote their products for local, regional, and international markets.

We would participate in several military displays with opening and closing ceremonies as part of providing entertainment to the public, culminating with an evening tattoo on the last day of the show. The army went to great lengths to remain visible in the public eye, and became a centrepiece of every such show. The highlight of the show was the weeklong one held in Nairobi. The chief guest, either the head of state or a visiting dignitary, gave the event a high profile.

As a member of the ASK organizing committee, I would liaise with the chairman, Major Hugh Collins, a retired British army officer who managed the society with remarkable efficiency. My role was to plan the participation of servicemen and servicewomen in static displays, and various field demonstrations such as dummy assault acts, casualty evacuation, and so on. These were aimed at providing entertainment to an anxious audience. Scripts and notes for such scenes with a general outline of the event for the announcer had to be prepared, and performances had to be rehearsed to ensure that the presentation taken to the public met with the highest of standards. Planning for the tattoo took extra effort as it required an expert input.

For this, we relied on professionals such as the organizers of the world-famous Edinburgh Tattoo, who were very helpful in providing assistance. The British High Commission went to great lengths to assist us.

The Kenya Army enjoyed a warm relationship with the public from its participation at these shows around the country and at many national events. The ever-popular marching song of the former KAR, "*Tu funge safari,*" was widely sung in Eastern and Central Africa, to where their origins are traced. The newly emergent armies continued with the same tune but varying lyrics that had a profound appeal because they were patriotic and stirred emotions. The spectators would often join in a singalong. The lyrics had a universal appeal and went like: "*We are the KAR soldiers; we are now starting our journey; pack up your bags, we are going on a journey; work of today is fierce; where are we going? Orders, on whose orders? The government orders. Whose orders? Orders of the commander, orders of the KAR; let's go and finish the enemy; then we return home and look after our families and cattle; we are ready, the KAR is ready, let's go.*"

SPORTS

The army encouraged participation in individual and team sports at the national level by giving generous time off to athletes who achieved a certain level of competency. Although Kenya has gained and maintained a global reputation for long-distance runners, its accomplishments in team sports such as soccer, cricket, and golf have been just as worthy.

I was co-opted into the national organizing committees of the Kenya Lawn Tennis Association, the Rugby Association, and the East African Safari Rally.

Lawn tennis was not a popular sport in the army, despite each base having excellent tennis courts. The expensive clothing and equipment were major factors behind why this great team sport was out of reach for many. I was able to seek assistance from the association who supported us by providing clothing, equipment, and instructors who gave their time *gratis* to run tennis camps. Some sports houses also supported us generously.

Slowly, the sport acquired interest, and we were able to field teams at various regional competitions. Despite our efforts, the sport remained low key and largely in the preserve of a few.

Rugby attracted marginal interest, but not the universal appeal we had thought it would. It was hard to offer an alternative to the traditional soccer, which had a universal appeal primarily because it was widely played and needed just a football and an abundance of enthusiasm.

Kids would play in bare feet in the street or in any open area by simply putting their clothing as goalpost markers.

Athletics, especially long-distance running, has put Kenya on the world map. The sport of running became popular by accident, since it evolved from the necessities of survival. Household chores, whether to fetch water from a river, take cattle for grazing, or pass messages about possible attacks from neighbouring tribes, to say nothing of having to return to dwellings before dark and rains, required running. Such a routine from which there was no escape built a natural flair for the activity and, over time, it formed into a habit, especially amongst those living in the highlands. We were able to attract a large number of these athletically inclined candidates and, as part of a national plan to encourage sports, we would give these athletes an almost preferential treatment with generous time off.

EAST AFRICAN CORONATION SAFARI

I was a member of the executive committee of the East African Safari Rally. This was a rally with participants driving standard motor vehicles across a distance of over 1,000 kilometres through Kenya, Uganda, and Tanzania (formerly Tanganyika). The rally had its origins in 1953, when it was organized to commemorate Queen Elizabeth's coronation, then called the East African Coronation Safari. My role was to coordinate the manning of various checkpoints where timings of participants were recorded and relayed to a control centre. These checkpoints, some of which were hidden, had to be manned to ensure that participants followed the route and did not take shortcuts that would disqualify them. We had to ensure that these remote checkpoints had reliable radio communication facilities.

The army along with the police assisted by providing mobile, vehicle-mounted wireless sets. Initially, participants were all local motor enthusiasts bearing their own costs. They would be assisted by their friends, some of whom would pre-position with an assortment of spares parts, tires, and refreshments, at various locations along the route.

The uniquely adventurous and demanding nature of the rally course in muddy conditions, along escarpments on narrow, slippery roads, rising from sea level to over 10,000 feet, from brutally hot to cold temperatures, was relayed to an enthusiastic global audience via television, print, and radio. As global interest in the event grew, major auto manufacturers got interested in using it to showcase and promote their newer models. They started sponsoring well-known rally drivers with generous prizes, and the event became part of World Rally Championships.

Welcoming Vice President Danial arap Moi at the start of the East African Safari Rally, 1968. Attorney General Charles Njonjo is on the left.

The rally was renamed the East African Safari Rally in 1960, and had its name changed again in 1974 to the Safari Rally. The rally put Kenya on the world map of motorsport, but a lack of volunteer support pushed it into organizational and financial problems. The situation was compounded with differing political ideologies in neighbouring Uganda and Tanzania, which resulted in it becoming a Kenya-only event. I was pleased to have played a part with the organizing committee for several years in its success.

With such participation, the army remained in the public eye and earned the respect of the common folks.

DEPLOYMENT IN GARISSA

In mid 1967, our battalion was due for a six-month rotational deployment in the NFD, based in Garissa. We embarked on preparations and conducted several rehearsals in field tactics, including laying ambushes and assaults on enemy encampments. We built physical endurance by going on long marches. We tried our best to familiarize ourselves with mine detection as the Shifta had started laying land mines in dirt roads to blow up vehicles.

In Garissa, our unit strength was similar to a battalion group, with detachments of a recce squadron, a field medical company, a squadron of engineers and signals, a detachment of ordnance, and a section of Kenya Air Force Beaver aircraft at our disposal. Our battalion headquarter was in Garissa, and we deployed company detachments to Wajir, Mandera, and Moyale. As the adjutant, I would represent my commanding officer at the provincial security committee meetings chaired by the provincial commissioner, James Mburu. We planned search-and-sweep operations in conjunction with the police, the paramilitary general service unit, and the administration police, to check on any illegal activity in and around towns and the villages of El Wak, Habaswein, etc.

The Ethiopians were facing similar secessionist threats in the Ogaden region bordering Somalia and to our north. Army headquarters directed us to initiate liaison with the Ethiopians and coordinate operations to neutralize the Shifta, who would escape into the other's territory when pursued by either party.

We mounted joint operations on two occasions with the Ethiopians to contain the Shifta. The Ethiopians were a pleasure to work with. The Ethiopian forces were led by General Gasese Ratta, governor of the neighbouring Boran province, who visited us in Garissa a few times and I met his operational staff at the border town of Moyale on a few occasions. He was the coordinator of our joint operational plans and executed operations.

With his seniority in rank and a much larger force assigned to our joint approach as a coalition, we had no problem working under him and happily took orders and implemented our side of the mission as best as we could.

The Ethiopians' intelligence network was extensive and admirable; it reached out into small villages and trading centres. The nomadic tribesmen who roamed freely with their cattle in search of grazing land and water passed information of Shifta movements to their handlers. Such information had to be acted upon fast, with the full knowledge that, by the time it reached the planning staff, it would already be perhaps a day old and, in a fast-moving nomadic lifestyle, late. The Ethiopians freely shared their intelligence with us and we were able to mount sweep-and-clear operations along our borders to drive the Shifta back into Somalia.

We achieved remarkable success by disrupting their command-and-control structure and capturing stockpiles of small arms, ammunition, and land mines.

General Gasese Ratta was an astute, combat-orientated soldier. He was a natural leader and at home with the soldiers, who adored his bravery. He pushed endurance to the limits, and yet displayed a deep humanitarian compassion for the soldier.

During our deployment in the NFD fighting the Shifta insurgency, we struggled to win the hearts and minds of the local population, in spite of our assurances that we were there for their safety. We were going to be there for only a short time until the Shifta menace was eradicated and stability restored. As part of our efforts, we sent our medical personnel to run clinics in villages with medicines to treat the sick. It is amazing how this small humanitarian gesture worked in our favour, as we would get intelligence from them and would then track those planning attacks.

We had to be careful while analyzing such intelligence, knowing that the villagers were playing a double game. They appeared loyal to us during the daytime, but at night were unable to refuse lending support and assistance to the Shifta, whom they probably knew since the Shifta hailed from their proximity or had family ties with those coming from Somalia.

The Shifta would depend on the villagers for shelter, food, medical care, and intelligence of our movements, not an uncommon situation in any similar theatre. Before reacting to intelligence reports, we had to ensure that they were not deliberately misleading to get us into an ambush.

With this knowledge and assumption, we turned the situation to our advantage. We discreetly developed a secret, small, but very effective intelligence-gathering network of our own spies who would be our eyes and ears in the villages. Flashes of Merti would come to mind. Such an activity was not part of our rules of engagement. We still went ahead, boldly ensuring that our initiative remained closely guarded within a very small number. Who was there to check on us? Our patrolling was aggressive, to ensure the relative safety of villagers and transport trucks delivering food and essential supplies into remote areas. We operated with creativity and improvisation. We sent our nightly sitreps and army headquarters was quite satisfied with our performance and results.

GARISSA, WATERMELON FARM

Quite accidentally and by sheer chance, an opportunity arose to win over the local population in an unorthodox manner. Of particular interest was the work of a missionary, Brother Mario from the Catholic Church, who was sent to open a church in Garissa with an assistant. In a predominantly Muslim culture, it was going to be tough to spread Christianity and attract a following. This was a generational task. One early morning, Brother Mario came to our tented camp, introduced himself, and explained that he had an idea to grow watermelon on the banks of the River Tana that went past Garissa. He needed our help and, during his brief visit, I found him a fascinating person with an engaging personality.

His appeal for assistance appeared genuine, and he left me no choice but to find ways to help him. We had no direction from army headquarters on how to handle such a situation.

We decided to provide low-key assistance as part of our training exercises, but to channel it through the provincial commissioner, thereby giving it legitimacy. Hopefully, it would be looked at positively by Nairobi. Our camp was by an airfield not too far from the river. His plan was that, by having a plantation, he would have to hire staff and later house them and, in the course with this captive audience, he would launch classes leading to the teachings of Christ.

In his thinking, NFD was a cut-off part of Kenya where the majority of people were of Islamic faith, closer to Somalia in ethnicity and culture. They felt removed from a predominantly Christian Kenya. River Tana, the longest river in Kenya, formed a natural division between the NFD and the rest of the country. NFD was sandy, arid, and sparse, with desert-like conditions. The western part, with relatively rich grazing lands, was predominantly Christian. To highlight this imaginary division, the locals in the NFD would refer to the lands west of River Tana as *huko Kenya*—"their Kenya"—and to the lands forming the NFD as *NFD yetu*—"our NFD." Such a division was mentally ingrained, and we would hear the locals describe it as such.

Brother Mario's was a visionary mind and had merits to complement the government's long-term goals for economic development. They were not meant to win hearts and minds only, but to provide much-needed economic incentives to create employment and introduce agriculture as a way of life. We never found out if he had permission from his superiors or even the government to initiate such a mammoth undertaking, or if it was the church's goal to spread Christianity. We accepted his umpteen requests for supplies of water, medical assistance, and, on rare occasions, transport, simply because it was our way to build bridges with the local growing community. We took it upon ourselves to assist.

At our weekly provincial security committee meetings, we would discuss him to ascertain if he had any subversive intentions.

We wanted to know the nature of his work and his mission, all along accepting that he was like any other missionary who lived a simple, basic life, and gave all he could to his people with good intentions; to us, he did not appear subversive. He gained our trust with his honesty, sincerity, openness, and never-ending appeals for assistance. We believed in his mission, and did whatever we could to assist him, remaining cautious that we had no authority to deviate from our own mission, which was to search and destroy the Shifta menace.

He had done his homework to satisfy that the conditions of soil, weather, and proximity to water were suitable for growing watermelon, and the project required very little education or dependency on fertilizers. He was confident that, in due course, other suitable crops would be introduced to the arid region, offering agriculture as a way of life for the residents.

His long-term goal in the region remained to offer a settled lifestyle that would allow residents to be engaged in an important economic activity, raise their families well, and send their children to school. In time, he was convinced, they would acquire skills and trades.

As such, they would earn a better living than that afforded by their nomadic lifestyle, where every day was a struggle to find watering holes and whatever little greenery was available for grazing cattle. Brother Mario's plan had merits, but we did not want to get too involved as we had no directives to assist with his noble initiatives. We supported him by providing water with our water truck which he would store in his two 500-gallon reservoirs. Beyond this and the few times when we sent our medics for minor emergencies, he was on his own. In the beginning, workers from around Garissa would commute to his farm to work; later, he built basic huts to house them on his farm.

In due course, we would see him and his staff digging shallow trenches into which he would pump water from the river for irrigation. His watermelon plantation was a huge success, and he would often bring melons to our camp.

In later years, I would hear that watermelons from Garissa were of a superior quality, and Kenya started exporting them. With this success enjoying widespread publicity, Garissa became popular with entrepreneurs, and many such farms mushroomed along the Tana River, fulfilling Brother Mario's dreams of creating employment, changing the lifestyle of local residents from nomadic to agricultural, and, in the course, contributing to the national economy. I did not follow further to see how his church attendance fared. The event signified an oft-heard term that it needs only one person with a vision to change the world, and *shape a destiny* for many.

We were relieved after our six-month deployment and returned to Langata Barracks. I returned to defence headquarters as an operation and training staff officer again in early 1967, continuing with my previous role. The insurgency was lifted in 1967 with the signing of the Arusha declaration in Tanzania. Soon after, Kenya signed a defence agreement with Ethiopia in 1969 as a deterrent to any other neighbour invading Kenya.

STAFF COLLEGE, CAMBERLEY, UK

As a captain, I was lucky to have won a spot at the British Army Staff College in Camberley for a year-long staff course in 1969. The attendance was made up of largely senior majors, with an average age of thirty-five. I was only twenty-five, and the youngest student on the course. I was promoted to major whilst at Camberley. This premium institution trains mid-career select officers with a demanding program for higher command. It was exhilarating to see the familiar names on the course attendant list: Captain Patrick Brooking of the Royal Dragoon Guards, an armour officer who had been my instructor at Sandhurst only five years earlier; Major Chuck Ivey, Duke of Wellington's Regiment; and Major Richard Corkran, Grenadier Guards, both of whom had served with me in Kenya with the King's African Rifles and the newly formed Kenya Army.

At Camberley, we all came with different experiences at different levels and this offered a great opportunity to learn from each other. The course was focussed on strategy and doctrine at a senior tri-service level.

A lot of time was spent in the library researching campaigns, studying commanders, and preparing debating notes to highlight their strengths and weaknesses. We would rotate frequently from syndicate to syndicate and work with different students on complex projects that required group input. This gave us an opportunity to build a strong team at short notice, a task that would prepare us well for later years.

Camberley showed us what we could make of ourselves given an opportunity to shape our future goals.

We had several prominent, high-level guest speakers, such as General Westmoreland (Westy) of Vietnam fame; Field Marshal Bernard Montgomery (Monty), who had previously been to Sandhurst to speak to us; and Lord Louis Mountbatten, First Lord of the Admiralty and the last Viceroy in India. Each one spoke on subjects that mattered to us: leadership, strategy, tactics, military industrial complexes, and the Western alliance's robust stand against the uncertainties of nuclear threats from a hostile Soviet Union.

I recall with fond memories a fireside chat with Mountbatten to which a small number of us were fortunate to have been invited. Sitting next to him was inspiring; he was a warm, friendly person but with a distinct privileged, aristocratic personality. He was interested to hear how the post-colonial Kenya was faring. He was just as interested to know the background of ten or so of us. He asked me of my lineage, where my ancestors in the Indian subcontinent came from, and what made me pursue a career in the army. Our chat was social rather than professional. I responded very briefly about the military heritage and warrior profile of my ancestors. I mentioned that my ancestral roots were in the village of Hoshiarpur in the Punjab. Upon hearing this, he immediately responded that that was the district where boys are raised with the martial spirit that provided the bulk of recruitment into the Indian army. He equated Hoshiarpur with Aldershot, home of the British Army. It was a remark that lifted my spirits to dizzying heights.

As part of our course curriculum, I had to submit a paper on a subject from a select list.

Superpower rivalry was testing nerves in Africa, the Middle East, Southeast Asia, Europe, and, most recently, Cuba, during the missile crisis. There was no shortage of recent theatres of confrontation to study as a term project. Preference was given to areas where we had operated and had first-hand experience highlighting successes achieved, obstacles encountered, and recommendations for future implementation. It was evident that such in-depth studies would eventually find their way into high-level planning committees in the hope they would add value for future contingencies.

Camberley, as with any high-level training institution, acted as a laboratory to evaluate the rights and wrongs of the past to improve planning for the future.

My term paper was based on my experience in participating as a senior officer, where I coordinated and managed a response during the Nyanza floods in Kenya in 1967. The title of my paper, "Military Aid to Civil Power," was well received. It was encouraging to hear from my mentors that, despite not following a textbook guide, it had originality and—with scarce resources but plentiful enthusiasm—enough drive and energy to restore social stability with improvisation. I was pleased to have been complimented by my mentor, Major General Mike Mathews of the Engineers Regiment. Mike was a classic officer and a gentleman, ever so helpful with a warm, friendly personality. He took pride in guiding us, often after work hours. On leaving Camberley, he was appointed engineer in chief of the British Army. With a desk job in Whitehall, London, that kept him busy, Mike still found time to welcome me warmly on my numerous visits. I was comfortable sharing the many personal challenges I was facing, and he always provided a keen, listening ear, and imparted valued advice and guidance. He had a long-lasting impact on me.

Four other students whom I befriended exposed me to how destiny shapes our future beyond what one plans. The year 1969 is mostly remembered for man's landing on the moon on 20th July. This was a historic event and our classroom lectures were cancelled to watch it unfold live on a black-and-white television in the crowded mess.

We all held our breath seeing Neil Armstrong take the first steps on the moon and saying those famous words, "One small step for man, one giant leap for mankind."

DESTINY, UNPLANNED; LIBYA

Another event awaited us, this one predating the moon landing. On 9th July 1969, we were participating in an exercise outdoors. Colonel Hassan from Libya was a student at Camberley, and we were in the same syndicate. A secretary interrupted to deliver a note to the directing staff who showed Hassan the note. He proceeded to give us impromptu bear hugs of the type that implied that we would not see each other for a long time, or perhaps never again. He was gone. A little curiosity surrounded this abrupt intrusion that had seen Hassan leave us midway through a presentation. We wondered why he was pulled out, but assumed that it had to be an important matter.

That evening, we heard on the evening news that a twenty-seven-year-old army officer, Muammar Gaddafi, had staged a coup in oil-rich Libya and deposed the monarch, King Idris. He had appointed Hassan as army chief and that was the reason Hassan had been abruptly recalled. Gaddafi had sent a Libyan Air Force plane to bring back Hassan. Like us all, Hassan lived in the single quarters of the main campus building. We would go for meals in the dining room together and, on weekends, to pubs. Hassan was a friendly person who liked socializing ahead of work demands. He would happily ask for assistance to complete his assignments so that he would not have to spend long hours in the library. Most of us, where we could, obliged, especially when he was part of our team; otherwise, his neglect or incomplete work would have reflected negatively on the team.

During all this time, he never gave us a hint of what lay in store for him. After he left, we did not keep in contact and didn't know what fate came his way. In such circumstances, it was not wise to befriend anyone who was implicated in a coup. Army-staged coups were becoming a regular occurrence in many countries in the 1960s, as the colonies were gaining independence.

In later years, we all saw how Gaddafi brutally ran the country and mercilessly eliminated confidantes on the slightest of suspicions, who at one time may have been close supporters of his. We never heard from Hassan again.

Ethiopia

Two officers from Ethiopia, Colonel Mesfin Gebre Kal and Lieutenant Colonel Alam Alamayu, lived next door to me in the single quarters. Mesfin was smart, astute, and in possession of an analytical mind. He participated in all activities, regardless of what it took to prepare for them. His contributions to syndicate discussions were wholesome and well presented. He would thoroughly research a project and it was evident that he wanted to be seen as a capable officer destined for higher command. He was a loyalist to Emperor Haile Selassie and, being amongst a small number of senior officers who had been selected to attend various mid-career courses abroad, it was only a matter of time before he rose to the top.

Alam was on the quieter side, but a great team player who contributed to a group assignment in whatever manner he could. He easily overcame any shortcomings with his charm, and it was always a pleasure to enjoy his company and sense of humour. On return to Ethiopia, both Mesfin and Alam rose in the ranks to brigadier. I met them on some occasions briefly when they paid separate visits to Kenya. We would exchange news about each other's families, but other than that, we did not maintain contact. When a young army officer, Mengistu Haile

Mariam, mounted a Soviet-backed coup in 1974 to depose Selassie, I lost track of them both.

Sudan

Colonel Abu Zaid Ibrahim Hilall from the Sudanese Air Force was a likeable, friendly colleague who was conscientious and studious, and had a deep love for aviation. He would engage in lengthy discussions on the capabilities and performances of various aircraft, describing their optimum use in combat. In syndicate discussions, he would always defend the aviation factor as a key component of warfare. It was a pleasure to listen to him, and he added depth to our many presentations. On return, he rose to command the Sudanese Air Force.

I met him in Khartoum after I retired, and he was always a gracious host. He showed concerns for the future as the fundamentalist Muslim Brotherhood was establishing a presence in Sudan with backing from Egypt and starting to draw a large following. As a senior military officer, he had to remain a loyalist to the government of the day, and we did not get into discussions of politics or of what lay ahead for him. I would hear later that he was at odds with the president, who made little effort to contain the spread of the Brotherhood with their extremism. I never heard from him again.

KENYA, COUP?

In post-colonial Africa in the early 1960s, a popular trend in the colonies was for the military to take over their country. Some elected officials in several countries went afoul of the army with their arrogance, corruption, and tribal animosities. In the eyes of others, they did not measure up to the exacting demands of leadership required to run a country. This required all the intricacies of administration, knowledge of economy, and means to keep a diversified ethnic population content. Suitable infrastructure-development programs that would lead to economic progress that benefited every community had to be spread out across the country. In some countries, only the loyalist tribes enjoyed such benefits to the neglect of others.

Coups had been staged by the military in many new countries that had recently gained independence. There were mild murmurings that Kenya may be the next. The events of early 1963 when the 11th Battalion of the KAR mutinied were still fresh. Neighbouring Uganda was showing signs of unrest, with Idi Amin making waves. Ghana fell to military rule. Neighbouring Somalia was in constant turmoil, with the army having taken over. Zaire was battling between rival military leaders. Several other West African countries were in military hands. In Kenya, tribal alliances were always shaky, as the newly elected government cherry-picked regions for roads and infrastructure development as a reward for loyalty and support. Some regions enjoyed development and prospered, but others remained left out, thereby fomenting unrest.

After returning from Camberley, I was posted to defence headquarters as the military assistant to the chief of defence staff, Major General Joseph Lele Ndolo, in 1970. I was responsible for preparing briefing notes for a host of tasks, including the budget, parliamentary questions, liaising with diplomatic corps, foreign troops training in Kenya, and visits and inspections of units and attending to Ndolo's personal commitments. Ndolo had risen from the ranks of the King's African Rifles and was a capable, likeable commander with a natural flair to inspire. His lack of formal education never bothered him. He had fine-tuned the fundamentals of leadership, morale, man-management, discipline, and teamwork from his early days as an NCO, a WO, and an effendi (a rank between a WO and a commissioned officer created by the colonials to command a platoon). He understood the basic, simple needs of a soldier: what it takes to satisfy concerns of pay, safety at home, care of his children, availability of food, care of his cattle, and the confidence that medical care would be there if and when it was needed.

Military Assistant to chief of defence staff, Maj General Ndolo, 1970.

Wedding Day, 23 August, 1970, Nairobi

These basic issues are key to fostering an esprit de corps, building a regimental spirit where a soldier would put the regiment's interest before his own. His grasp of tactics he'd practiced and experienced in Burma and against the Mau Mau was amazing. For a person who enlisted with no formal education and only determination and a resolve to succeed, he did very well, and was an example of what can be achieved given an opportunity.

Coming from the warrior Kamba tribe, he was looked upon as their voice, and enjoyed a hero's status wherever he went. He liked partying and would regularly entertain friends and government officials at his huge farm in Sultan Hamud, 100 kilometres east of Nairobi in the heart of Wakamba country.

I was often at such gatherings and, as his MA, was involved in organizing many of them, a task I accepted as part of my duty.

Ndolo was a distinguished combat experienced officer who'd served in Burma in William Slim's 14th Army with distinction and bravery. He was a people person—always jovial, considerate, and compassionate.

He attended several social events as a chief guest to support whatever reasonable cause he had been invited to. These causes would be to raise funds to extend a classroom or buy equipment, or participate in a *harambee* to raise funds to bore wells in a village or any other cause benefitting a community in Kenya.

As such, his popularity within the army slowly extended to the larger public, and the press would cover his travels extensively. He became a media star and the ruling elite, largely Kikuyu surrounding Jomo Kenyatta, started becoming suspicious of a potential challenger to Kenyatta. They were concerned about whether he had ambitions to mount a coup and take over the government. This was far from reality. In my observations, he was a soldier's soldier throughout. His only escape from soldiering was to manage cattle on his farm in Sultan Hamud through a manager. This is the farthest I would imagine him venturing outside of the army to organize and run something, but not a country. I find it hard to make any connection between him and the purported coup.

My personal relationship with Ndolo was warm, friendly, and professional. Despite his lack of a good command of English, he more than compensated for it with his abundant enthusiasm, leadership, energy, and openness. During my closeness with him, I never suspected that any of his actions could be deemed treasonable. He was not tribalistic in his actions, and was at ease with whomever he was dealing. He had every opportunity to put his own tribal loyalists in key positions, but did not interfere with the promotion boards and allowed them to make recommendations for promotions or key appointments as they felt appropriate and on merit.

I never heard him speak against the government, point accusatory fingers at senior bureaucrats or politicians, or entertain phone calls or visitors who could be described as undesirables or suspicious. I was his gatekeeper and right-hand man who managed his demanding schedules and ran the office with utmost professionalism.

The political situation in the country had become tense with the assassinations of outspoken politicians Pia Gama Pinto and Tom Mboya, an upcoming, young, and well-liked politician from the Luo tribe. Kenyatta's deputy, a Luo from the Western region, Jaramogi Oginga Odinga, had been detained in 1969 for questionable activities rumoured to be treasonable and this led to riots in Kisumu. Odinga, commonly known as "Double O," had an affinity for the Communists, especially the Chinese. He even wore collarless jackets similar to what the Chinese wore during the Mao Tse-tung era to express his affinity with a Communist ideology. Suspecting possible unrest, the government clamped down on public gatherings and monitored the activities of suspects closely. In neighbouring Uganda, Idi Amin, a former colleague of Ndolo when both served in the King's African Rifles, ousted the president, Milton Obote, in a coup that sent shockwaves across East Africa. Kenya was ripe with rumours of potential instability.

On one incident at his farm in mid 1971, Ndolo had invited a large number of guests, most of whom were of his Wakamba tribe. The senior-most were Kitili Mwendwa, the chief justice, and another prominent politician who was a member of parliament from his neighbourhood and a fellow Kamba, Gideon Mutiso. I was part of the team that organized a barbecue, called *nyama choma*, or roast meat, a common and popular social activity.

The Kikuyu chief of staff Colonel Lucas Matu and another senior officer, also a Kikuyu, Colonel Stanley Kamau Ayubu, drove with me to the farm. The outdoor barbecue had some 100 guests and the party continued late into the night with traditional dances, *ngoma,* provided by locals for entertainment. Lucas Matu, Stanley Ayubu, and I left in the early evening when the party was still in full swing. The intelligence services picked up some rumours that Ndolo was planning a coup to take over the country and that this gathering of key close associates was to develop a strategy and finalize plans to execute. The news spread fast, and it took many of us who had been at the gathering by surprise.

Rumours gained momentum that Kitili Mwendwa was to remain the chief justice and, when a coup announcement was made, he would swear in Ndolo as the new president.

The following morning, I noticed that Ndolo looked worried and did not want the customary morning briefing, choosing instead to be left alone in his office with the door closed. That very morning, Kenyatta gave an emotional, nationalistic speech focussing on loyalty and castigating coup planners without naming individuals. The atmosphere at Ulinzi House where defence headquarters was based was tense. We heard Kenyatta's speech on the radio where it interrupted regular programming and was accompanied by nationalistic songs to stir up emotions and win support.

We knew that something major was wrong. Looking out the window from the eleventh floor of Harambee House, we saw a large presence of the paramilitary unit, the General Service Unit (GSU), which was organized, equipped, and trained as an infantry battalion, but came under the police department. It was widely believed but never discussed that the GSU was a counterweight to the army. There was almost no joint training between the GSU and the army, nor any closeness of any type. Both were kept apart, and the GSU's camouflage clothing resembled an army unit's clothing. It was also assumed but never spoken that the key security institutions were headed by loyalist Kikuyus from the Kikuyu heartland of Kiambu and Nyeri, with Bernard Hinga heading the police, Ben Gethi the GSU and intelligence service, and James Kanyotu the special branch. I enjoyed a personal, warm friendship with all three, and they all would extend generous social courtesies in their office after a meeting but never at their mess or at home. Bernard Hinga and I would play squash regularly at the police sports club and, afterward, would retire to the club bar for a drink. Bernard played good squash and always gave me the run-around.

Ndolo remained on the phone, making several calls using his private line for quite a while.

Late that morning, Ndolo called me into the office and showed me a short, two-paragraph letter he had just drafted that he was going to hand-deliver to Kenyatta by noon. He asked me to proofread the text, type it, and not discuss its contents with anyone.

It was a sad and difficult task to type Ndolo's resignation letter. Several of us army officers who were at that gathering in Sultan Hamud feared that, by association, we would be implicated in a potential coup. We suspected our careers would come to a meteoric, unhappy, crashing end with more fears of landing in court. Time passed, with much stress at work and sleepless nights, as each day seemed our last in uniform. Fortunately, we did not have to face any accusations of wrongdoing or interrogation for disloyalty. We remained alert, knowing that our telephones were tapped, our every move monitored by the ultra-efficient intelligence branch of the police, the special branch.

We probably got paranoid that a secret network of spies was amongst us, and we had to be very careful of what we said and what company we kept. It was common knowledge that the special branch had its spies embedded in the army, and it was of utmost importance that we remain careful of what we spoke. Taking sides with any party or pointing accusatory fingers at anyone for whatever reason was not wise. Flashes of how I had set up a secret intelligence-gathering network at Merti would haunt me since it could be assumed that the same had been repeated here, as well.

A late midday special bulletin on the radio announced that Ndolo had resigned. Calls by placard-waving protesters called for the immediate resignation of Chief Justice Mwendwa. Ndolo was gracious to some of us who were close to him at this final juncture in his life. A few of us who had offices on the eleventh floor, numbering about twenty, gathered around my office to wait for Ndolo to emerge from his office. He gave us a bear hug, shook our hands, and wished us well.

Seeing Ndolo leave the office and walk alone along the corridor toward the elevator, hearing echoes of his steps, with just a briefcase, a look of sadness, and a teary face after nearly thirty-six years of loyal service, was heartbreaking.

With these circumstances surrounding him, he did not get any customary dining-out farewells in the officers' mess or inspections of retirement guards of honour, or any of the other accolades that might recognize his long service. He had pursued an exceptional, colourful career and risen from an illiterate private to become the chief of defence staff in the rank of a general. He had encounters with history at many junctures, but ended with much sadness, despair, and loneliness. *Events shaping destiny?*

After he left and retired to his farm, I maintained contact with him as I treasured his friendship but was aware that the special branch was following him closely and that my close association with him, my lack of knowledge of the coup plan notwithstanding, may put me in danger. Ndolo died in a car accident alone close to his farm, an event that aroused further curiosity around its circumstances, thus ending his life with even a sadder twist.

Mwendwa resigned shortly thereafter, and Justice Surinder Sachdeva, a senior magistrate and family friend of ours, convicted ten conspirators. They had been implicated in planning a coup, some of whom readily pleaded guilty and got lengthy jail time. Following these convictions, several rallies were held across the country to re-affirm loyalty to Kenyatta and the government.

DEFENCE HEADQUARTERS— STAFF DUTIES

The new chief of defence staff was Lieutenant General Jackson Mulinge, another Kamba with a distinguished service record.

The relationship between Ndolo and Mulinge was professional, but it was evident that it was not close.

They had a similar start in the KAR, but perhaps rivalries for promotion came in the way when the Africanization program sped up. Mulinge was always friendly to me, but we never became close. I was appointed a senior staff officer in charge of operations and training still at defence headquarters with additional responsibilities of liaison with the diplomatic corps, a lieutenant colonel appointment. I held the post for two years, earning the salary and perks of the rank, but was not promoted.

I have often wondered if my closeness with Ndolo negatively impacted my future career prospects, since most of my peers with much less combat and staff experience and some with questionable credentials went on to senior ranks. In later years, most of my peers would retire at brigadier or general rank. Such a development caused me much discomfort, and I started wondering if this was an indication of what lay in store for me.

Despite this, I continued to perform with the same dedication and loyalty without showing any bitterness. During the early 1970s, we continued to maintain close defence consultations with our traditional

Western allies: Britain, Canada, Germany, France, the United States, and Israel.

Regionally, we maintained a close defence liaison with Ethiopia, who appointed Colonel Tariku Nigatu as its first military advisor who was popular and remained in that role for over twenty years. India sent its first military attaché, Colonel Mohinder Chatwal, to Ethiopia, from where he covered Kenya. Canada's defence liaison was handled by the first high commissioner, Ms. Margaret Meagher, with a staff of two, later handled by her successor, Murray Cook. We had regular meetings with the diplomatic corps and developed a warm, friendly relationship with their defence advisors.

The need for training continued to increase. As old equipment was gradually replaced with new and additional equipment procured, our requirement for specialist training increased.

The need to train officer cadets at a faster pace became a priority, as we had to fill newly created command and staff appointments with qualified personnel. Officers had to be hired with specific professional skills such as medicine, legal, engineering, and accountancy. We started recruiting such trained professionals and, as an incentive, gave them a starting rank of captain followed by a specially designed, scaled-down, regimental boot-camp course at Lanet. We made sure that we were not too hard on them for fear of their quitting midway. Since our facilities at Lanet catered primarily to the basic boot-camp and rudimentary officer cadet training, we initiated a closeness with Nairobi University to assist us. The armed forces were expanding at a rapid pace, and we needed professional support from wherever we could get it.

With limited finances for defence and security, Kenya executed various defence agreements with friendly countries under which defence equipment was supplied, often as a gift or on generous terms. The donors recognized Kenya's strategic location, along with its growing economy and massive economic potential for the future. They undertook to bear the costs of training personnel not only for the armed forces, but extended it to offer places at universities for other disciplines.

There were indeed trade-offs, handled at a high ministerial level, to which we were not privy. We would hear that Kenya granted lucrative commercial concessions from agriculture, fishing, mining, oil exploration, infrastructure development, and so on for such generosity. We prepared our shopping list indicating the range of the equipment, services, and training we required, and followed through, making the most of whatever was offered.

Our biggest supporter was Britain for historical reasons. With English as the language of communication, a similar legal code, and a similar civil service organization, continuation of the British style of government was the preferred choice, and so was adopted.

We liaised and coordinated numerous military arrangements where Britain had executed defence agreements. Under these agreements, Britain maintained a small liaison detachment, British Army Training Team Kenya (BATTKEN), to coordinate the training schedules of British Army units who trained in northern Kenya in desert warfare and around Mount Kenya in jungle warfare. BATTKEN also assisted us in many ways from time to time, and we had a wonderful working relationship. As a result of my regular work with them, the British High Commission placed me and a few other staff officers on their guest list for several social events. On one such event in 1971, it was a privilege for Aruna and I to meet Prince Charles and Princess Anne at the residence of the high commissioner, Eric Norris, in Muthaiga.

With Aruna in her military style El Al Israel Airlines uniform, 1970.

WRITE A BOOK?

My closeness with the British High Commission continued with successive defence advisors and high commissioners. Colonel Brian Tayleur, a cavalry officer from the 14/20th Lancers, had arrived as the defence advisor, and my regular liaisons with him led to meeting the high commissioner, Anthony Duff, to whom I took an instant liking, a feeling that was happily mutual. Our relationship evolved slowly, and we would play squash at the exclusive Muthaiga Club whenever our schedules allowed. He would invite Aruna and me to numerous social events at his majestic residence in the leafy, exclusive suburb of Muthaiga.

As military officers, we operated under strict guidelines when socializing with members of the diplomatic community, aware that wherever we went and whatever we did, our movements would be monitored by the special branch. We had clear instructions not to overly befriend members of the diplomatic community and had to seek approval each time an invitation arrived. The biggest concern was not with the invitations we received from our traditional Western allies, with Britain topping the list, but with the more questionable, discreet invitations some of us would receive unexpectedly from Communist embassies.

We were amidst a superpower quest for influence in the region, and Kenya was part of this growing Cold War rivalry.

There was something extraordinary about Anthony Duff. He displayed an impressive, elegant personality that effused tact, polished grooming, and social finesse that was in keeping with diplomatic tact. With his warm, friendly character, he was charismatic and, despite his high status, he handled me as a junior young officer with utmost respect and cordiality. Could that be because I had given him a run-around in the squash court? I don't know, but this would hardly be the reason.

However, I began admiring him and, in hindsight, I see he was one of the idols to whom I looked to for inspiration. After our squash game, we would retire to the men's bar for a drink. If we went outside on the patio, we would sit on a specially designated table that the club had reserved for him, regardless of whether he was around or not. Nobody was to be allocated that table.

As our closeness grew, he would tell entertaining stories of where and when he'd participated in WW II with immense pride. His exploits were legendary—full of daring adventure and extraordinary bravery. He came from a distinguished naval family and his father had been an admiral. He never bragged about his short naval career to appear a hero, but when relaxed, he gave some insights of hair-raising suspense about where he'd played a hero's part.

He would narrate one particular event with meticulous detail, describing a time when, as a young naval officer in the rank of a lieutenant commanding a wartime submarine, he nearly drowned. Whilst patrolling off Norway, he came under heavy German fire and the submarine was disabled. It landed on the sea bed in a horizontal position through his sheer ingenuity and presence of mind.

With seconds ticking as the submarine was descending, he used the crew as a human ballast to shift weight such that the sub maintained its horizontal profile and, miraculously, he managed to get it to rise to the surface with no casualties.

What he did was a demonstration of unimaginable bravery on the sea bed. He disconnected the compressors which provided breathing air to the crew and diverted it to the sub's main compressor. That is what got the sub to rise to surface instantly. This was a remarkable feat of leadership that displayed not only bravery but initiative and creativity.

For this action, he was recognized with a Distinguished Service Order. In his words, his greatest relief and joy was that he did not lose any of his crew—an officer and gentleman all along.

I always looked forward to a beer and chat after squash, as his wartime experiences were thought-provoking, fascinating, and close to what one would see in a movie. He was just a natural entertainer who held court in whatever company he found himself. He was just as interested to hear about my past: where my ancestors came from, how and where I grew up, and what made me join the King's African Rifles. Much to my pleasant relief and disbelief, he would impress me with his knowledge of my martial ancestry in Punjab, which provided the bulk of personnel for the British Indian Army during British Raj.

He was a well-read person. He was taken aback by my escapades to Egypt to visit the battlefields of El Alamein, my hitchhiking in Europe, my trip in a trawler to Iceland, and the role I played during the Shifta emergency. He was the first person to suggest that, sometime, I should consider writing a book, as the journey I had covered at a young age was fascinating and would be an inspiration to others.

Sadly, I did not give his advice a second thought. But with constant pressure and encouragement from many, I followed through on his suggestion nearly forty years later and, hence, this book!

After Anthony Duff left Nairobi in 1975, I lost contact with him. But, much to my surprise, he called me a few years later when he was passing through Nairobi from Salisbury, Rhodesia—present-day Harare, Zimbabwe—where he had been appointed deputy governor. His task was to see a smooth transition of the country from Rhodesia to Zimbabwe, which became an independent country in 1980.

I was simply overcome with emotions when I heard that it was the deputy governor, a very senior officer, calling me at home to invite me to meet him for a drink at the Muthaiga Club, his watering hole of the past. This was hardly the type of invitation you would sit on and consider whether it conflicted with other commitments—my acceptance was immediate, much like a response to a command! By now I had retired from the army, and it was gracious of him to offer himself as a reference in my job search, should I ever need one, and assistance if I chose to settle in the UK.

Anthony Duff would then go on to become head of Britain's spy agency, the MI5, where he handled reorganization of some sort at MI6 and government communications headquarters (GCHQ). I had been to Bletchley Park, Cheltenham, outside London, where GCHQ was located as part of a trip from Camberley, and so knew how well protected and secretive the place was. This was the place where the German Enigma and Lorenz cipher codes were broken in 1942 by the famous genius, a thirty-year-old Alan Turing, with his team of mathematical wizards. This was the intelligence-gathering nerve centre of Britain. From here, worldwide communications of all kinds beyond those of security interest were monitored, intercepted, and, probably, broken into with deceptive messages inserted as diversionary tactics.

A game of cloak and dagger played to the fullest. During our trip, we were only taken to the briefing centre and visited a few offices and shown where the codes of the famous German Enigma machines had been broken during WW II.

Our colour-coded entry passes clearly marked where we were allowed to go, and we had to display these tags on our lapels for identification. We were strictly forbidden from deviating from an assigned group escort, wandering off, or taking notes, photographs, or any kind of recording of any discussion or briefing. We were in a high-security location and our every move and step were closely monitored.

Anthony Duff had a profound, lifelong impact on me and I have always considered how lucky I have been to have known him. After retirement, although I met him very infrequently, I treasured his company, wisdom, and advice beyond being entertained by his stories and admiring his remarkable life. It was not the longevity of our interaction that had a profound impact on me, but the depth of his personality and the interest he took in me, despite the huge distance between our social and professional statuses. I always looked forward to meeting him during my numerous short transits via London after I had immigrated to Canada, but I did not succeed in meeting him whenever I wanted to. My schedule was always tight, with priorities to build a new life in a new country, and I hardly expected him to be available to meet me for a mere courtesy call. He was surrounded by priorities of a national and global significance.

Given the discreet but high-profile job he was in, it was understandable that he was unapproachable. Getting to him required clearances at several layers of security and a demonstration of justifiable reasons to meet. Mere social courtesies, regardless of how brief, were therefore out.

I succeeded only once in meeting him for a lengthy cup of tea some time in the late 1980s at the Cumberland Hotel by Marble Arch in central London, by which time he had retired. He looked in great health, blending with common street folks with his modest attire.

His appearance was hardly reflective of a man who had played a key role in the corridors of power in London during the waning years of the Empire and had once served as a close confidant of Prime Minister Margaret Thatcher. Our conversation was stimulating and he would tell me how fulfilling his new occupation in retirement was: serving the homeless of London. This was by no means a fall from grace, but a self-imposed desire to leave a fulfilling legacy by helping the underprivileged who had dropped out of society for whatever reason, though certainly not a reason of their choosing. He was a remarkable individual who had a dignified character and grace with a deep passion to help others throughout. Before parting, he wished me success in Canada and asked if I had written my book—a *second reminder!*

POST-ARMY

From the mid 1960s, the racial situation in Kenya was becoming tense. Over a period, the Indians had established themselves as successful businesspeople. With sheer hard work, they ran small shops in remote areas and developed manufacturing and processing plants in larger cities. They created employment opportunities that offered a wide range of trades and skills. Their contribution to the national economy was huge. Their success was reflected in outward displays of affluence: a large, well-kept house, newer cars, and frequent trips abroad. A close communal network provided assistance and support to one another in several ways. We would hear that even a need for financial support was made discreetly and kept within the community. Such credits and loans were offered and secured only on trust, since a default would render the individual untrustworthy with no chance for future assistance. The result was that almost all the businesses were in Indian hands and the agriculture sector was in white European hands.

In 1971, I was posted to the 7th Battalion at Langata Barracks as company commander of Charlie Company. I was familiar with the unit from my days as the adjutant, and my transition was smooth. I concentrated on training, since our annual unit inspection was three months away. We were to be judged on our shooting skills, performance in field craft, forced endurance marches, tactical exercises (with us operating in both offensive and defensive roles), ceremonial parade drills, and unit administration (covering accounts, armoury, medical, signals, and vehicle detachments). We were at liberty to plan our training schedules. The inspection was spread over three days and at company level the results were a straight pass or fail. Our performance, along with that of other companies, contributed to the overall battalion's operational readiness. We had gained recent experience in the NFD and at various ceremonial performances in Nairobi. These held us well and, as expected, we were rated "operational above average," and a celebratory party followed.

Company Commander, Charlie Company, 7th Battalion Kenya Rifles, Langata, 1972.

I started thinking of retiring from the army in 1973 for a variety of reasons.

The main reason was that, on a couple of times, my juniors with much less experience and with merit taking a secondary role, were promoted over me. At first, I took it that fair assessment had resulted in the best candidate being promoted, but soon a pattern emerged in which promotions and key appointments were an extension of the government's Africanization policy to reflect a balanced tribal representation. There were only a handful of commissioned officers of Indian and European heritage—fewer than ten—in the armed forces who, through their professionalism, loyalty, and merit, had landed in key positions.

The tribal factor in itself was becoming a confrontationist issue on a national scale within the civil service and government agencies. Regions with existing infrastructure such as agriculture, health care, schools, and roads where manufacturing plants had been set up enjoyed a boost, with the consequent economic benefits of low unemployment and thriving trade. Other regions that were either arid or not economically hospitable to development were left out, creating an imbalance of opportunities. The ever-widening gap among the tribes with their insulated ways of life and the growing economic disparity compounded the social climate with ancestral rivalries surfacing. The rich were getting richer and the poor were getting poorer.

With such a volatile social climate where race defined and determined opportunities, a large number of Indians and Europeans, feeling uncertain about the future, started looking for other countries to migrate to. The Indians in Africa were, by now, second or third generation, well-established, and calling Kenya home, as were others in Uganda, Tanzania, Zambia, Zimbabwe, and beyond. In all these countries, race became an issue of colonial divide-and-rule policies. In fairness, the racial imbalance in employment and businesses had to be rectified. The government approach was justified, with legislation setting quotas in employment that trumped race over merit. The Indians were widely visible on a day-to-day basis as they ran shops and businesses all over the country. They became an easy target.

Some politicians, with their rhetoric outbursts to gain attention, stirred emotions at rallies by implying that the Indians were enriching themselves at the expense of the locals. This was hardly a direction to maintain social stability.

Since Indians ran most high-street businesses, they were the visible face of a "foreigner and success." Some radical politicians directed their rhetoric and outbursts at these shopkeepers, saying that they were taking away wealth from the native Africans. The Indians were also well established in the manufacturing and service sectors. The Indians became an easy prey, as were the Europeans who owned and ran large estates. The government undertook to redress the economic disparity by initiating strictly enforced racial quotas. The government designated some high-street shops in most towns to have an African partner followed by the manufacturing and service sectors.

It was only a matter of time before they, too, were forced to take on Africans as partners, regardless of their talent or suitability. Some Indians became creative by simply taking on Africans as token partners so that they could renew their business licenses, but kept management and control of business in their hands. This was a difficult situation to redress past imbalances and create equal opportunities, and a good deal of time had to pass before a truly multiracial society would emerge.

With legislation entrenched to speed Africanization, I felt that I did not have a secure future, especially for my young children, in Kenya. Emotionally charged events in the region with clear racial overtones had created fear and uncertainty. The next generation would face numerous racially charged obstacles to overcome beyond high school since the only university, the Nairobi University, was already burdened with eligible applicants who could not get a place. A population explosion in Kenya that increased from seven million at independence in 1963 to over twelve million a decade later (and over forty million today - 2016) meant that getting admission to the only university would be difficult, if not impossible. Even for graduates of overseas institutions, if we could afford to send our youngsters there, would face difficulty getting a job on their return.

The recent events in neighbouring Uganda, where a former King's African Rifles soldier, Idi Amin, had toppled Milton Obote in a coup and thrown out the Indians on impulse, sent shivers.

Amin gave the Indians ninety days to get out of the country with no compensation, despite the fact they were second-generation Ugandan citizens. Their only crime was that they were of Indian ancestry. Added to this was a universal racially charged climate in the region that endorsed throwing Indians out, as they were still labelled foreigners. Their economic successes were equated with those who were trying to steal from the native Africans. A fear of uncertainty for the future dominated many a discussion.

For me, two incidents stand out in memory. The first was when the Indians were thrown out by Amin, a trainload of them passed through Nairobi with a brief stopover. They were on their way to Mombasa, where they would board ships to return to India. India had pledged to accept Indians who had ancestral ties to the country.

As the train pulled into Nairobi, its cabin doors remained closed and the occupants were not allowed out. Police patrols on the platform ensured that no one exited or boarded the train. Word had gotten out for the community to arrive with prepared care packages with food, blankets, and clothing, which would be passed into the desperate hands stretching out from the train windows. These care packages were given freely to whomever reached for them in the hope that they would share their contents with others in the train. The scene mirrored the trainloads of Jews that were bundled by the Nazis into crowded trains on their way to the gas chambers in Europe during WW II.

The train pulled out, its windows filled with the desperate, crying faces of people who had been forced to abandon all their possessions and life savings in a land they had called home. Now facing an uncertain future even in the land of their ancestors, they were empty-handed, filled with loneliness, and bereft of plans. It was emotional and struck a reality of sadness.

About the same time, with Africanization enforced through legislation in government employment, massive restrictions around where and what type of business an Indian could run came into play, and sent shockwaves.

The Indians felt as if they were unwanted foreigners who did not belong to the country. Europeans and Indians started leaving Kenya voluntarily for other countries to start life afresh. Anticipating a wave of immigrants, Britain tightened its immigration policies and passed legislation that gave a time limit to accepting immigrants from the former colonies who still held British passports.

The clock was ticking here, as well, and a large number of Indians left for Britain in the hope of starting a new chapter in their life. Other Western countries such as Canada, the United States, Australia, and many others in Europe opened their doors to these Indians who had developed a reputation for entrepreneurship, hard work, and success. The Indians were now in demand for their skills. Some countries fast-tracked their applications to immigrate with generous incentives of subsidized travel and concessionary loans to start businesses. Canada stood amongst the leaders by accepting mass migration.

I was limited in my options to seek employment and that, too, had to be temporary until I decided what to do and where to go. I had no business background and, since my father had been my family's only breadwinner, earning a nominal civil servant's salary, I did not have the luxury of walking into a running business operation or inheriting one. Neither had I accumulated large-enough financial savings to give me a head start. My search was limited to seeking employment, hopefully at a level that was as senior as that to which I had been accustomed.

With Aruna, Meera and Sanjay, prior to retiring from the Army, 1974.

On retiring from the army in 1974 prematurely mid-career from a promising early start with ambitions and high expectations, I was confident that I would make it anywhere with determination and courage. The army, with its generous perks, provided every comfort including housing, transportation, and medical care, and that had left us with no incentive to build a nest egg for an eventual retirement sometime in the future. The only exception was Brigadier Miles Fitzalan Howard's advice whilst welcoming us as newly minted young officers where he compelled us to start saving through deductions made at source every month.

Unlike many, few of us never deviated outside our profession in pursuit of a business venture to build an asset or accumulate wealth for eventual retirement, which seemed a long way away. We were professional soldiers, and focussed all our energies on advancing the profession of arms and our careers. Our salary was modest, and we managed our affairs quite well within our resources. I had no idea how to run a commercial enterprise. The only exposure I had to business was running an officers' mess, an enterprise that required a basic knowledge of accounts, inventory control, and the management of a small staff. My only option remained to seek a well-paying job at a senior-enough level.

GHOSTS—BELIEVE OR NOT

As a priority, I had to find a place to live, and so rented a newly constructed double-storied, three-bedroom townhouse, a *maisonette*, off Lenana Road in the Kilimani suburb of Nairobi. The bedrooms were upstairs with a family room, dining room, and kitchen on the main ground level. My elderly parents were living with us. The place was conveniently located to downtown and the nearby Alliance Francaise French school, which my children attended.

One evening, I slept in the children's bedroom to give them company as they were not well, nursing a mild fever and cold. Late at night, probably midnight, I was woken up by a beautiful young black lady, elegantly dressed, adorned with jewellery, and smiling. She tapped me on the shoulder and said, "Come, come," upon which, without a second thought, I got out of bed and started following her to the hallway.

There was a metal door with glass panels at the far end which she simply walked through and disappeared. I saw this, and realized that she could not have gone through the door without opening it, as it was closed, latched, and locked. Something was not right. I got scared, abandoned going back to the children's bedroom, and returned to our bedroom, quietly sliding into the bed and covering my head.

Aruna woke, sensing something was not right. I quietly whispered to her that a stranger had been in the house and was now gone. She got scared and I started sweating, still tucked in the relative safety of the quilt. I wondered where my army bravado went as I abandoned my two young sick children in the adjoining room at the very time they needed me! Very early the following morning, I shared the unusual experience with my mother, who said that she had also seen an elegant, well-dressed black woman matching my description a few times, but had not told me as I was already under stress to relocate from a secure army job to an uncertain future.

On sharing this episode with our neighbour's guard, who had lived there for a long time, we were told that, apparently, the murder of an attractive, middle-aged woman had been committed several years earlier, and that her spirits had been visiting the house. Aside from us, even neighbours had heard and seen the spirit coming in the form of a ghost that matched my description.

My parents approached a priest to ward off evils and, whether it was superstition or not, we conducted daily prayers at home to cleanse the house.

After two weeks, on one occasion as the prayers were coming to an end, we heard a loud swish noise, much like something rushing out, and the priest confirmed that the spirit had been driven away and that the house was cleansed. Regardless of this assurance, we started looking to relocate. We stayed in the maisonette for another six months before moving into a house close by that we had purchased. We never saw or heard the spirit again. After moving out, out of curiosity, I would visit the maisonette and the adjoining house only to hear that nobody had seen the spirit again. Our landlord pleaded with us not to make the news public so he would not lose on his investment. Having read about ghosts and the never-ending debate around their existence, I can say that they do exist.

BURGLARY—HOUSE BROKEN IN

When we moved into our new house, another surprise awaited us. This was a double-storied house with bedrooms upstairs. I had returned from a long business trip and was having difficulty sleeping. Jet lag, sleepless nights, and long working days were taking a toll. It was a hot evening and we had opened the windows for a cool breeze on the main level and upstairs. We did not have air conditioning, and alarm systems and window security bars had not become standard features in houses. I decided to catch up with my reports, and worked late into night at our dining table on the main level. Probably past midnight, I put my working papers and files into my briefcase and took it upstairs to our bedroom. I placed the briefcase by my bedside to be readily available just in case I remembered to check up on anything when I got up. I fell asleep.

Early the following morning, our domestic hand, Mbevi, frantically knocked on the door and said that, on his way, he'd picked up my briefcase and some papers that were lying by a laneway, some 100 metres away. We were dumbfounded. On investigating, we discovered muddy footmarks from the porch upstairs leading to the bathroom and around our bed up to behind a heavy curtain, some three feet from my bedside. I pieced together my detective work and realized this is what had happened: Sometime by or past midnight, a barefoot intruder had gained access to our bedroom from the connecting bathroom. He must have gotten in via the open bathroom window by climbing onto the porch roof. He then gently walked into our bedroom and hid behind the long floor-to-ceiling heavy drapes when he heard me walking up the staircase. He waited quietly until I was asleep, picked up my briefcase, and left via the way he came in.

No harm was done to anybody, but the outcome could have been devastating had I opened another window in the bedroom and faced the intruder. In such a circumstance, it is fair to assume that an intruder would be armed, perhaps with a gun or knife. Either way, had a physical confrontation resulted, he would have been on a brutal offensive and, after attacking me and perhaps attacking or kidnapping a family member, he would have escaped.

As with any such a terrifying situation, one naturally feels vulnerable when his private space is invaded and the mind races to unimaginable scenarios: what if I had confronted him as I came up? What if he had been surprised by a family member and what fate would have come their way? What if I had sensed noise, breathing, or unfamiliar movement in the room? Was he alone or did he have lookouts outside? How did he work his way around so well? How did he time his entry?

Had he been at home before? Was he a contractor—a plumber, an electrician, or handyman—whom we may have engaged? From his side, it was an act of bravery to have attempted such a risky intrusion. We strengthened security at home as my business travels would take me away frequently for long periods.

GENERAL MANAGER, *OF A LOCAL ENTERPRISE*

I would soon find out that my previously unchallenged command authority would be diluted with endless negotiations and a lot of give and take to reach a compromise. The first was in the tourist industry, where the owners of a travel enterprise had plans to build new game lodges in national parks. It was clear that they wanted me aboard since, as a newly retired officer, I had close contacts in the government and would be useful for tackling bureaucratic hurdles. The remuneration package included an option to become a partner sometime in the future, and it looked promising, though it was a long time away.

Businesses were tightly controlled by a handful of Indian families who created cartels and monopolies that an outsider could simply not break into. They controlled raw materials, manufacturing, and the distribution of their products to retail stores. They did very well and accumulated a lot of wealth, despite lacking in formal education. Most of these businesses were now in their second or third generations, headed by individuals who had left school midway and acquired skills on the job. They were going to be my paymasters, and it was not up to me to question their credentials or social values.

I applied for a job in a manufacturing company that specialized in textiles, from spinning cotton to weaving cloth.

I chose the textile company where the pay was good and the title of a general manager suited me. The title was uplifting, as it was a continuation of providing leadership and managing assets, which I assumed would be straightforward. This enterprise, with a manufacturing plant on the outskirts of Nairobi, had 500 employees and I equated it to a battalion strength. I was new to the world of a thriving commercial enterprise whose fate was now in my hands.

I had to learn the skills of this specialized trade fast. In my first week, I immersed myself in learning all about the business psyche and the manufacturing processes, and, more importantly, to meeting supervisors, technicians, sales managers, the accountant, and my plant manager, and having them brief me on their roles. Mastering new vocabulary such as yarn, warp, weave, rolls, and the technicalities of an assembly-line plant was going to make this an interesting job.

In the second week, a shop steward accompanied by a union representative walked into my office with a notice that, unless pay and benefits were improved, the employees would go on strike. We were given two weeks to respond. This was an eye-opener in the world of union negotiations. The clock started ticking and it was ticking fast. I had hardly been there long enough to understand what it took to run a plant, to familiarize myself with assembly-line procedures, to learn how raw material was ordered such that a supply chain was set for just-in-time deliveries. I was still learning how a line of credit with the bank functioned, how to handle a salesforce, and how to set targets. I had to prepare for mandatory inspections by government inspectors, enforce compliance with numerous regulatory bodies, and attend to employee welfare concerns and numerous other inescapable chores. And here I was facing a major dilemma of a possible plant closure within two weeks of taking a new job. My day would start at 0500 hrs and I rarely got home, fifty kilometres away, before late night, seven days a week. The union issue began to bother me, and I had no clue what to do.

On impulse, I initiated a meeting with the shop stewards two days later. I heard their never-ending demands for increasing wages, and improving benefits and working conditions.

Some of the demands were genuine and had been ignored, including the provision of transport to take the sick to a medical clinic some ten kilometres away. For such an absence, an employee's wages would be deducted for time away from work. They asked for re-assignment of some supervisors who did not get along well with employees. They asked for time off for weekends as, until then, the plant functioned seven days a week and the shifts rarely started or finished on time. They asked for time off for midday and mid-afternoon tea breaks, the extension of lunch time from twenty minutes to an hour, and the dedication of a room where employees could have their tea or meals, like a canteen. Until then, employees would have their meals outside and, if it rained, they would huddle either under a huge shady mugumo tree or assemble in the crowded garage.

Church attendance on Sundays for most employees was a must. Since some worked on Sunday, they could not attend a service with their families. Flashes of spirituality as a bond for high morale reminded me of its significance. Earlier teachings of morale and welfare flashed in my mind. The biggest issues were wage increases and leave benefits, which were beyond my authority since they had to be weighed against a matrix of various costs: production quotas, direct and indirect overhead costs, and revenues generated.

I was not familiar and knew nothing of such detailed cost breakdowns. This was an accountant's job. I also wondered if the plant was overstaffed and if there was duplication of work, a situation that would require streamlining efficiencies and would most likely lead to layoffs, which would trigger another confrontation with the union. Only two weeks into this new job, I took on the challenge, and presented the situation to the shareholders with recommendations. I was relieved that they were understanding and gave me a leeway to negotiate, since I believe they had faced such threats in the past and had pretty well ignored them.

At my second meeting with the shop steward and union representatives, I addressed each and every concern, maintaining all along that we could only commit to what we could afford and that we had limits.

I reminded them that we had to improve on efficiency and production targets, which were below industry standards. This meant an overhaul of the entire workforce to see if there was duplication of effort and, if so, there would be layoffs.

I drew a clear distinction between the roles of management and employees, emphasizing that it was the shareholders who would dictate terms and conditions of employment. These were to be enforced by the management, complying with the laws of the day and the understanding that unions were not in control but obliged to work in harmony with management. I gave a mild threat that, should the unions insist on their unaffordable demands, the shareholders had the option to close down the plant with massive layoffs. They may have seen this as a strong-arm tactic, but it was a reasonable approach. There had to be a fair give and take to achieve a good working relationship. Regardless of our differing positions, we were part of a team and it was upon us to make it work.

Much to my relief and satisfaction, the shop steward and union representative were understanding, and took it upon themselves to convince the employees that we were being fair and to accept the improvements as offered. The employees got an increase in wages, shift timings were regulated to specific hours, and the plant closed for the weekends to allow for maintenance work on machinery and for employees to attend church.

We arranged with the neighbouring medical clinic to send a medic to the plant twice a week to attend to the sick, and reassigned supervisors. The implementation of changes was to be immediate. A crisis had thus been averted, much to the relief of everyone.

As a further step to building a closer relationship between management and employees, I initiated a monthly collective meeting, a *baraza,* after work hours on the last Friday evening. I would meet them with my three managers, and employees could ask any questions or present any concerns they had. The object was to build a bridge in our relationship.

Such a gathering had never been held before and the first one had almost a carnival atmosphere to it, with rambling questions. With no end in sight, it lasted some two hours. Subsequent *barazas* became more structured, with shop stewards presenting a list of questions in advance, which gave us an opportunity to prepare responses. The exercise got us closer to the employees and weakened the shop stewards' stranglehold on them. *Barazas,* which would last a convenient thirty minutes to an hour, were now a routine.

The improvements had a profound effect: morale blossomed, absenteeism declined, and, much to my and shareholders' joy, the production manager confirmed that production figures had increased and equipment breakdowns had declined. Overall, the plant became efficient and company profits soared.

TRAVELLING SURPRISES, YEMEN

Some three months later, as part of an expansion program to go into a joint venture overseas, I was sent to the Port City of Hodeida, Yemen, on the Red Sea to negotiate a joint venture. This was with the owner of a textile manufacturing plant, Sheikh Mohammed Ibrahim, who was also a respected community leader and elder. He was in his early fifties. The plan was for us to provide technicians, upgrade his plant, arrange lines of credit for raw materials, and receive compensation by a flat fee with an option to buy shares in his plant. This was another situation of a novice landing amidst seasoned, experienced business people, and pretty well not knowing how to proceed. But facing this challenge of immense proportions, I put on a brave face, accepted the assignment, and arrived in Hodeida via Aden and Sanaa.

I was pleasantly surprised that my initial meeting with the sheikh at his manufacturing plant was warm, friendly, and promising. He displayed an eagerness to proceed with his expansion plans at the earliest. He asked for a technician to accompany me on the next visit a month later who would identify the technical requirements of our proposed joint venture. We both impressed each other with promising expectations of a fruitful relationship. He invited me for lunch at his residence the following day.

The noon-hour lunch was a feast, held on the second floor of his mansion to which he had invited a large gathering numbering around twenty. At lunch, I was introduced as his foreign guest and, by local customs, was now accepted as a guest of his clan. The bulk of the conversation at the two tables was in Arabic, which I did not understand, but my assigned interpreter, Majid, kept me abreast with his running softly spoken commentary. After lunch, we retired to the open-top third floor, customarily used in the region much like a deck for socializing owing to the hot weather. A canopy provided shade from the blisteringly hot sun. We sat on a carpeted floor with large pillows to support our backs. I sat opposite the sheikh on the far wall. A *hookah*—a traditional smoking pipe—was passed around and, in spite of the fact that I was a non-smoker, I took a puff on Majid's whispered advice that I would otherwise be seen as a snob.

The setting appeared to have some formality and everyone seemed to know where they would sit. There was no rush to sit next to the sheikh, a move that would have implied a closeness or attempts to win favours. Sure enough, this was a court of elders where the sheikh handled minor breaches and infractions within the clan that were brought to his notice and, as a clan judge, rendered verdicts.

Three accused were brought in one at a time by a male escort of about the same age without any handcuffs or force and both remained standing. The "prosecutor" spoke out the infraction loudly and the elder clan members, acting much as a jury, gave character references on the accused.

The defendant then gave his side of the story. The accused was allowed to give his version of the infraction. The sheikh would there and then ask the elders for their opinion on an appropriate reprimand or punishment. He rendered judgement, which was like sending the accused to be a slave to another family for two to four days. This was speedy justice in a clan system that existed as part of a well-defined hierarchical chain of command that thereby enforced total submission by members to their elders.

On a signal from the sheikh, Majid and I left as the hour-long proceedings were coming to an end. On my way down, on the first floor, hearing female chattering voices in the kitchen, I slightly opened the drawstring curtain, saw beautiful faces and uttered the word,

"Shukran"—thank you. The women quickly covered their faces with the burka veil and silence followed. A short while later, Majid came down and said that he would pick me early the following day for my flight back to Sanaa and Aden.

I was puzzled by this, given that I had four more days to continue with our negotiations. Apparently, the sheikh found out in those few moments that, as a foreigner, I had trespassed into his private space and saw the faces of his wives without his permission—a catastrophic breach of protocol.

The sheikh did not want to see me ever again, and had instructed Majid to have me sent back at the earliest.

At the hotel, there were armed guards by the main entrance, on every floor, and in the elevator. I was terrified. Being in a strange country, not speaking the language, not knowing anyone, and having gone afoul of a well-respected elder and successful businessman, I was left with no option but to remain calm and respectful till my departure. Any thoughts of going to the hotel cafeteria for dinner, or taking an evening walk in the neighbourhood or to the market for souvenir shopping were out.

I did not sleep that night and, if I did, I would be woken by the guards' thumping boot steps. It was a long night. Majid came very early at 0500 hrs to drive me to the airport. He was polite and courteous, talking very little. As he dropped me, he gave me a traditional bear hug, wished me well, and drove away.

I took the domestic Alyemda flight to Aden, connecting with a Pakistan international flight and arriving in Nairobi at noon. I was relieved that I was back home safe. The failed unfortunate business side of my visit did not bother me, since in my mind, I had done nothing wrong. When I unexpectedly arrived in the office the following morning, I was uncertain how I would be received. When I explained what had happened and the circumstances leading to my premature return, my boss was just relieved that no harm had been done to me. He said that perhaps it was for the better, since a similar fate may have come to any of our staff who were certain to visit Hodeida in follow-up visits. It was a lesson learned to be very careful of cultural matters of all kinds.

KENYA, FUTURE

Kenya remains a bulwark of stability and free enterprise, and a hub of economic growth in the region.

The Kenya Defence Forces have acquired a reputation for skill and efficiency, both on the parade ground and in the field. Since independence, the armed forces have grown enormously. The engineer battalion, under the umbrella of training, has contributed to national development by constructing bridges, building roads, and sinking bore holes in arid areas to make drinking water available to villagers. It has responded to natural disasters such as floods, drought, and famine by providing support and medical care. On the security issues, it has maintained stability by close support to the police services.

On the international stage, since the early 1990s, Kenya has earned respect in a variety of United Nations and African Union-sponsored peacekeeping roles in Angola, Namibia, Mozambique, Southern Sudan, Democratic Republic of Congo, Somalia, Sierra Leone, Macedonia, Croatia, and East Timor. Kenya contributed troops and observers in these missions. Its biggest success was in October 2011, when it launched the first armed campaign outside its borders to drive the terrorist organization, Al Shabab, out of Kismayu in Somalia. Al Shabab until then was posing a threat to Kenya's tourist industry with its attacks on hotels at the coast to create panic. The armed forces gained invaluable experience in sustaining an effective command-and-control capacity and operational and logistic support to much success.

Kenya has focussed on building training facilities for the armed forces to prepare its officers to international standards. The National Defence College was set up in 1997. Here, domestic, economic, foreign, and defence issues are discussed to develop national policies. Exchange students and staff from many countries provide an international environment to discuss complex issues. On the logistics side, emphasis is also on building homegrown skills by expanding various technical services and repair workshops where equipment, aircraft, and naval vessels are serviced locally.

In recent years, with ongoing conflicts in the Great Lakes Region and Eastern Africa, a need to train officers in multidisciplinary work to plan and execute peacekeeping operations has come up. This was handled with a great deal of pride, with Kenya setting up the International Peace Support Training Centre in Nairobi.

Here, personnel from several countries attend courses to evaluate a wide range of conflicts and prepare officers to engage in law enforcement. Such preparation also introduced the workings of joint operations, where officers and men from other countries participate. The centre provides research, training, and an educational forum where seconded personnel from donor countries, the African Union, and the United Nations assist with financial support.

CANADA—A LAND OF MILK AND HONEY

In 1976, I started thinking of migrating. I could not figure out if it would be to the UK or Canada. Thoughts of uprooting the family and re-establishing in a new country presented innumerable concerns. This was a big decision, and I was hesitant to take the plunge without a clear thought. Aruna hesitatingly agreed, but rightly showed fears around the prospect of our not making it out. She was bringing in a paycheque from her employment, first with Ethiopian Airlines and later with El Al Israel Airlines. We were managing ourselves reasonably well. My paycheque was there, but I did not feel secure for the future, especially for our young children, Meera and Sanjay. I constantly felt the urge to discuss this major decision with whomever I had confidence in for independent advice just to satisfy myself that our decision to migrate was a sound one.

Whenever I discussed my future intentions, several senior officials who knew me well and in whom I confided supported me. Brigadier Miles Fitzalan Howard, my first brigade commander with 70th Brigade King's African Rifles, was one such officer on whom I would pay courtesy calls when passing through London. I would meet him at either his Arundel Castle in Sussex or his London townhouse in Euston Square. He encouraged me to move to the UK, offering to line up a job for me with a government agency. By now, I was considering Canada, where I had been on several business visits and liked. My visits had been in August, and the weather seemed perfect, even tropical-like. With Canada ranking tops in my choice to migrate to, it required a lot of courage to decline the offer from the well-respected Miles Fitzalan Howard.

Around the same time, the Canadian high commissioner in Nairobi, Murray Cook, encouraged me to move to Canada on the grounds that I was already known by several Canadian companies, some of whom had offered me lucrative jobs. He continued that Canada was a young, growing country in which a lot of opportunities lay ahead. I started thinking of making my future home in Canada and never wavered.

Over the years since retiring from the army, I had established a vast network of contacts with senior officials of several companies. I approached them for assistance; they offered opportunities from employment to working with them on a retainer and commission basis. The terms were lucrative. Either option required lengthy travels abroad to drum up business, and this suited me perfectly. Making presentations to clients at decision-making levels was a continuation of my days in uniform, and it did not mean lowering my standards or self-esteem. I was prepared to learn the intricacies of a vast range of products in support of my marketing efforts. The decision was made. We moved to Montreal, and lived in the suburbs of Saint Bruno on the South Shore, a small bedroom community of about 25,000, divided equally between the English and the French.

On migrating in 1980 and on sheer impulse, I started thinking about starting my own business. A few Canadian companies I had contacted appointed me as their agent to promote their goods and services to several countries in Africa. By now, I had mastered a little of what it took to have a business of one's own. Aruna took a job in a managerial capacity with Air India and, later, with SITA, the Societé International Telecommunications Aeronautics, in Montreal. With her salary, we did not have to dip into our savings.

STARTING A BUSINESS, SELF-EMPLOYED

Since representing a company to promote their products did not require a financial outlay, I thought this was a safe option to pursue.

I became self-employed, marketing the products of the several companies I represented to Africa and Central Asia. The companies with whom I affiliated myself manufactured or provided specialist services of a wide nature. This included pre-fabricated housing, de-salination plants, hydro power generation, defence equipment, railroad equipment, air traffic control systems, veterinary products, call-centre services, real estate development, and property management. This was a diversified range that gave me the comfort to spread my net wide enough such that at least some would be successful.

As I transitioned into this business and took on various representations, I had to undergo an extensive learning process. The initiative had to come from me, as this was now my own business. I immersed myself in learning about the products and services of each principal so that I would be well prepared to promote them. This was a prerequisite to earn the confidence and trust of the client. I had to assure them that I knew what I was offering and was familiar with the generalities of each product. This was going to be a real challenge, requiring a lot of study and support from the principals, and there were no shortcuts.

For any sales pitch, I had to know what I was talking about as I would be addressing seasoned experts who would engage me in discussions. I had to know what the competition was, since that was my biggest challenge to counter. The client would most certainly conduct a detailed analysis of quality, pricing, delivery, sales support, and payment terms, and would compare my proposal with those of the competitors. A sound knowledge would be the basis to equip me for my calls on whomever I contacted and for whatever product I offered. It was critical to know and understand what I was selling.

I had to find the client and promote my principals. My remuneration was based on the amount of business I was able to generate. Unlike in a job where a paycheque at month's end is assured, I had to generate revenue, and this meant going out and drumming up business. If I did not do that, there would be no income. Period. I took the plunge into the unknown and there was no going back. I reminded myself that, if David Dobie could make it, so could I. My partial comfort lay with my wife Aruna's paycheque, which paid for groceries and ran the household. Some moments were emotionally draining like when, at the dinner table, the kids would ask, "Dad, are we going to be OK?" It required a great deal of courage to put on a brave face and hide any sense of uncertainty.

I knew I had limits. But this was no different than commanding a unit where I was expected to have a general knowledge of the technical units under my command but did not have to know the finer details. The need was to know the operational aspects of a product, and their benefits and limitations, but not the detailed technicalities.

The analogy would be that of a watch: to use it as a tool to read time without knowing how the many parts functioned. Or knowing how the operational limit of a Beaver aircraft with its range, load, surveillance capacity with the length of a remote airfield required for landing and take-off could be incorporated into an operational plan, rather than knowing how the aircraft was made. With this confidence, I knew it was up to me to learn on the job and to rely on my principals to train me.

My assumption was that the principals would identify other complementary qualities in me and would make up for my shortcomings with some training. I had to keep a positive attitude throughout.

I knew I would have a short time to make a presentation to a potential client and that there would be no second chance. I had to earn the trust and confidence of my future clients during this narrow window, and assure them that my principals would stand by their product. Name recognition was just as important—hence, my interest in selecting companies that had respectable reputations, such as General Electric, Rolls-Royce, Spar Aerospace, and several others.

With suppliers of defence equipment, I had no problem, as I was familiar with the range of basic equipment I had handled. But I discussed other factors such as packaging, shipments, delivery schedules, and payment terms in generalities, with the approval still left to the principals. I took on a wide range of representations, from hydro power generation to pre-fabricated housing, to manufacturing veterinary products.

My first representation was of a hydro power-generation system. I spent a few days at the company's plant in Quebec to understand the various technicalities of a turbine, generator, transmission lines and towers, transformers, and how they all interacted with each other. This was followed by learning the administrative aspects of bilateral and international financing. Such a mega-project with a value of over a billion dollars had funding provided by several institutions: the World Bank, Britain's Export Credit Guarantee Department, the Canadian International Development Agency, the Canadian Commercial Corporation, and Export Development Canada. I would be part of a team to negotiate the supply, installation, and commissioning of the power-generation system and simultaneously conduct discussions with clients in host countries for fulfilling financial terms and other contractual conditions.

Such a major negotiation spread over various levels required knowledge of infra-structural financing, loans, and credit handled at a ministerial level on a government-to-government basis. There was inevitable nervousness with such a level of complexity, and I was perhaps the only one in the team with limited knowledge on all fronts. The experienced team members who were vastly knowledgeable in various aspects of such a mega-project were most helpful, and guided and accepted me as a team player. They corrected me when I got into strange territory, which was often, but I guess they must have also recognized that I was able to compensate for my technical shortcomings with an ability to open many doors to facilitate smooth contractual negotiations. The negotiations took two years, and the project implementation three years.

The project was handled with remarkable efficiency, was financially lucrative, and reinforced my confidence that I could pretty well handle any product. This was education gained in the school of hard knocks, and it encouraged me to seek out other principals and build a portfolio. I started contacting other principals.

My next representation was marketing pre-fabricated shelter systems and the territory assigned to me was Africa—the entire continent! I selected five countries and concentrated on developing markets there. Here, again, I spent a few days at the company's plant in Calgary to learn every aspect of the shelter system, including manufacturing, transportation, erecting, and modifying them for specific uses. I had to study and understand the municipal building codes as they applied in Canada and hoped there would be some commonality in other countries. Compliance with regulatory authorities was necessary to ensure that permits, where required, were secured. My marketing efforts led to much success and, at one time, I was managing a crew of five over a three-year period, training local crews to carry out regular maintenance on the structure systems we had sold.

By sheer coincidence, I got involved with a company in Quebec in the veterinary field that was a subsidiary of a major European conglomerate. They had recently set up manufacturing and marketing operations in Montreal to expand their market into North and South America.

We won a contract to supply veterinary medications to a major client. I got interested in this new exotic line of products for which I had no product knowledge. But I enjoyed working with both the suppliers and the clients, as people on both sides were very friendly and engaging. The clients interested me more, since interacting with them required visiting farms, which I loved as it got me outdoors. I extended my interest to encourage my principals to outsource the manufacture of a few of these products.

Such a trend was gaining momentum for principals to keep their manufacturing costs down since they did not have to worry about maintaining expensive equipment, skilled highly paid staff, or expensive in-house laboratories. For the established marketing companies who had in-house manufacturing plants and established clientele, preference was for just-in-time deliveries, which eliminated warehousing needs and the limited shelf life of their products.

I handled the outsourced manufacturing of a small number of highly regulated antibiotics for animals and, in the course, learned the technicalities on the job. The outsourced manufacturing plant in Toronto had to be FDA-compliant. Such a certification was important to ensure every batch of medicines met quality standards and was safe and effective. Their in-house laboratory was useful to test samples of raw materials and finished products on site to ensure that the raw materials were pure and consistent with their declared percentage of ingredients.

Some minor changes at the plant were necessary to make the manufacturing process more robust and reliable. A strictly enforced system of controls at various stages of manufacture eliminated contamination, mix-ups, mistakes, and errors. The plant was subject to random, unannounced inspections by the regulatory branch of Health Canada. A thorough, well-defined manufacturing process, commonly called the CGMP—*Current Good Manufacturing Process*—was rigorously implemented, thereby assuring that products met quality standards. The in-house laboratory for periodic sample testing was a key element of quality control since a consumer cannot detect deficiency in a product from touch, smell, or sight. The project lasted over ten years to much success.

I took on several other representations and participated in government tenders to supply clients with a range of products, from specialized tires for heavy mining and construction vehicles, to call centres, air-traffic-control systems, supplies of patrol boats for law enforcement, and the repair and overhaul of aircraft engines.

As my marketing business grew, I immersed myself in property management which I could handle when I was not travelling and this required simple administration. I built up a small portfolio to manage commercial strip plazas and later invested in real estate. My business as a one-man operator grew rapidly and, since then, I have never looked back.

I was off to a very happy and successful start in 1982. As a testament to the great opportunities that lay ahead, and as a newcomer to the land of hope and opportunity, it was a great honour to have been selected to organize and manage Canada's participation at an environmental exhibition in Nairobi, in which several Canadian companies participated. The exhibition was held in a park in downtown Nairobi. Prime Minister Pierre Trudeau attended the United Nations-sponsored exhibition, and it was my task to welcome him, *as a Canadian,* and escort him around the various booths. This was an honour of immense proportions.

Welcoming Prime Minister Pierre Elliot Trudeau at the Canadian stand at a UN sponsored Environment Exhibition, Nairobi, Kenya, 1982.

This was the first time I met Trudeau. He had a reputation for high intellect and was a star on the global stage. It was awe-inspiring to be in his company in the full glare of the media. During my hour-long time with him, he lived up to his reputation for having a photographic memory and enjoying an extraordinary gift of being able to immerse himself in the intricacies of a subject instantly. He would ask serious questions on a range of solar-power, wind-power, and hydro-generation issues, and about the viability of the shelter systems that I was promoting, as if he was an expert. At each booth, he did not allow things to be rushed, and gave the participants the comfort that he was there just for them—a seasoned politician every step of the way. At a reception hosted by the Canadian High Commission one evening, he asked me how the exhibition had fared, again confirming his remarkable gifts for remembering people and events. I was overjoyed to be part of a gathering at such a high level and became convinced that this was only the beginning of a new life in Canada and that greater goals and accomplishments lay ahead.

Kanwal Sethi

More opportunities—travelling surprises, Kyrgyzstan

I knew Africa well but had no knowledge of Central Asia other than reading that after the break-up of the Soviet Union, many small, ethnically homogeneous countries with huge mineral deposits and in geographically strategic locations had emerged. They were in a hurry to get closer to the West. These new countries needed advanced technology to extract and develop their vast resources.

The world of international travel is never short of new experiences. And I never seem to be short of adventurous surprises. In late 1982, a requirement to upgrade the airport control tower arose in Bishkek, Kyrgyzstan, in Central Asia. I was representing a well-established Canadian company who had developed state-of-the-art technology in aviation with a speciality in air traffic control systems.

I arrived in Almaty, Kazakhstan, and was met by a former colleague, Rudy Rodrigues. We made the three-hour drive to Bishkek, Kyrgyzstan, where he lived. My meetings with airport officials and government bureaucrats were two days later. Rudy was the head of the World Food Program. He lived in a *dacha*, a huge, two-story bungalow set in the woods on the outskirts of Bishkek. *Dachas* were like big cottages, and were the preserve of the powerful and well-connected during the Soviet era, where bureaucrats would retire for the weekend to relax. After a warm welcome, he showed me around the house as he was leaving that evening for a meeting outside Bishkek.

The bedrooms on the first floor overlooked a covered verandah which was around the house. The wrap-around verandah provided some soundproofing from outside noises. At the kitchen pantry, Rudy pulled a rug that was covering a hidden hatch leading to the basement.

The basement was a bomb shelter. It was constructed with thick concrete, windowless walls and motion-operated lights. With these features, it qualified as a safe room.

There was only one item of furniture there, a large metal box that contained survival gear: a satellite phone ready to use with a battery life of four hours and two spare batteries as a back-up, water bottles, morphine with an injectable kit, flashlights, vacuum-packed food packs for a week's supply for four, four thermal blankets, and a battery-operated walkie-talkie radio set with direct links to the World Food Program office in Bishkek and the local police station.

On one end of an empty bookshelf, a button opened a secret door which led to a narrow corridor that I was told connected to an adjoining diplomatic house. I was mortified just to see the extensive steps available in case of an emergency, and started wondering about the status of the previous occupant.

He had to be very important and extremely scared to have such extreme measures in place for his safety and survival—something one sees only in James Bond-type suspense thrillers.

Before leaving, Rudy told me that I should secure the main, huge, wooden double door by placing a heavy metal bar across it, and should not answer the phone or open the door to anyone. He would be back the following day by noon. As he left the house, I secured it, as instructed. Soon thereafter, I heard some small stones hitting the outside verandah window, loud knocks on the door, and the telephone ringing.

I did not respond as per instructions, but got very scared, being the only occupant in a huge *dacha*, set in forested surroundings, in a strange place, and at night. Squeaky floors did not give much peace or comfort either. I fell asleep only to be woken up again very early in the morning by loud banging on the door. I recognized the voice: it was Rudy! Apparently, his meeting had been cancelled. He could not get a hotel room and ended up spending the night in his car outside with the security guard keeping a watch over him.

My following few days in Bishkek, which is set in beautiful mountainous surroundings, were uneventful. The country had recently emerged from the Soviet Union when it was dismantled, and the Communist methods of a command economy were evident everywhere.

Going into a government building was a challenge; as a foreigner, I had to get a permit from the ministry of foreign affairs, take it to the police department, who would check my credentials and stamp it, and then take it to the ministry of aviation, who would allocate a time slot to see the relevant officials.

The procedure would normally take two to three full days, half of which would simply be spent waiting in rooms to be called. I would question the reasons I embarked on such a burdensome mission, but gave myself the comfort that anyone contemplating doing business there would have to endure a similarly time-consuming schedule just to get an appointment with the right officials, with no guarantee of securing a contract. If I was not ahead of the pack, others would take over—*the-early-bird-catching-the-worm reasoning.*

Once I was at the meeting, the display of warmth and courtesy I enjoyed was in sharp contrast to all the inconveniences to which I'd been subjected earlier. The officials spoke excellent English and knew well what they needed. More importantly, they had a good knowledge of what was available in the West, especially Canada. My handicap was that I was not technically conversant with a major project of such a scale which required the evaluation of specifications and performances of various pieces of equipment.

Carrying colourful brochures as part of my presentation was not enough. We concluded our brief meeting on the understanding that a technical team should follow up, which sadly never happened. This was disappointing, since many other lucrative projects were to follow based on the success of refurbishing the air-traffic-control tower.

I would hear later that a few Canadian companies got involved in mining there and were very successful. *A lost opportunity for me. . . .*

Nigeria—more surprises

I was getting accustomed to long-distance international travel for business. The world of international marketing for which travelling to different countries is a pre-requisite is never short of strange experiences. I went to Nigeria in mid 1985 to promote Canadian-made, pre-fabricated shelter systems to the military. My presentations at the army logistics base on the outskirts of the port city of Lagos were promising, and a contract appeared to be in sight. With encouraging endorsements, I was to continue with contractual negotiations with senior government officials in Abuja, in central Nigeria.

Numerous mega-construction projects were proceeding at a rapid pace to develop the infrastructure, as Abuja was going to be the future capital of this huge country of nearly a hundred million people, the largest in Africa.

I arrived at the airport before sunrise on a weekday and, after check-in formalities with no baggage to check, proceeded to the departure gate and waited patiently in high humidity. As the flight number was announced for our flight to Abuja, we had to walk over to the aircraft, escorted and directed by the Nigerian Airways staff. Our aircraft was parked some 100 metres away. All of a sudden, we started running toward it. The faster few boarded the staircase and took our assigned seats. Soon thereafter, the cabin was filled with a crowd and passengers were arguing and claiming seats that had already been occupied. Those seated would not get up and those standing became argumentative. The confrontation was building up fast for a possible physical fight.

Such a chaotic scene was at every seat. Passengers seated and passengers standing had boarding cards with the same row and seat number. We were to hear later that the flight had been overbooked three times.

We could hear a further commotion from the overhead speakers that had not been turned off. The captain and the station manager were arguing about authority, about whose decision on whether to fly was final. The situation became chaotic and resembled a scene in a vegetable market, with loud voices drowning any conversation, and shouting and threats until police arrived and forced us all out. The aircraft doors were closed. Someone had placed four bags as markers around the aircraft.

The station manager lined us all up outside of these markers and said that, when he blows the whistle, we were all to run twice around the aircraft outside of the four markers. He would then open the aircraft doors and those who made it first would get to fly, with the remaining passengers having a choice for a later flight or a day-long bus ride. Bags already checked would not be off-loaded and passengers could pick them in Abuja at the airport when they got there.

The penalty for short-circuiting and not going outside of the markers was disqualification. I was one of the first to get in and secure a seat; fortunately, it was now open seating, and just as well.

I was pleased that travelling in running shoes meant not only comfort, but had saved the day. My daily regular running schedule had been a blessing. I did not know how the elderly and little children fared. Sadly, there was no bonus or upgrade to first or business class as a reward for being the first to enter the aircraft!

I had a difficult time sharing my experience with many who would just not believe my strange adventure. Much to my relief, the news item appeared in the *London Times* on September 26, 1985, with the story narrated by Tony Samstag.

Zambia—travelling surprises continue

On another trip to central Africa to promote pre-fabricated shelter systems, I took a taxi one very early morning when it was still dark from the intercontinental hotel in Lusaka for the airport. I had developed a habit of travelling light, with mix-and-match clothing that could be washed in the hotel room at night for a clean, presentable turnout for the meetings the following day.

The drive was perhaps fifteen kilometres, and there was dense fog with reduced visibility. About halfway to the airport, the car stalled and came to a stop. The driver and I got out. The early dawn breeze was cool, and we heard birds and dogs in the distance. There was not a soul in sight, nor any sign of any activity, and no other vehicles passed from whom we could have sought help. The driver, who obviously knew the area well, told me that he had run out of fuel but that the good news was that there was a gas station two kilometres down the road.

He cooled my nerves by assuring me that I would not miss my flight. His rescue plan was to remain in the driver's seat and for me to get out and push, maintaining that it was downhill to the gas station and pushing a Datsun four-cylinder vehicle would not be difficult. A lot of worrisome thoughts flashed through my mind.

Did I leave the hotel too early? Should I have waited for the hotel shuttle an hour later? We were stranded on a major road from the city to the airport and why would there be no other vehicle on the road? Would traffic build up shortly? Seconds and minutes counted. I became uncomfortable, and started suspecting that this was a trap. I did not want to remain outside in the event that he took off with my carry-on bag still in the trunk.

I was not concerned about my passport, airline ticket, and wallet as I had them on me, secured in a hidden traveller's belt under my shirt. It was important to remain calm and hide my frustration and fear. To remain friendly, I offered to switch roles, saying that I had a bad back and that I would sit in the driver's seat with him pushing the taxi. He willingly agreed.

A few minutes later, we arrived at the gas station, woke the attendant who lived upstairs, and, as we were filling, two minivans heading to the airport passed us. The fog was still heavy, but the day was slowly opening up. I was relieved that the taxi driver had genuinely forgotten to fill up and did not have any criminal intentions. He dropped me at the airport, profusely apologizing for the inconvenience. The saga was not yet over.

At check-in, I had to pay US$10 departure tax in cash or equivalent GBP5, but not the local Zambian Kwacha. The smallest US-denomination bill I had was $50; I'd used the other smaller bills for my taxi ride. The check-in Air Zambia stewardess gave me change in Kwacha, since she did not carry US bills—or perhaps she wished that I would leave the change with her.

At the security examination booth, I was told that I could not take out Zambian Kwacha in excess of 100 Kwacha, the equivalent of US$10. I was given a choice: I could leave my Kwacha change at the security desk or, sensing that I was not going to follow his suggestion for a "tip," buy bars of soap from a kiosk across the hall, since Kwacha was not acceptable in the duty-free stores in the departure lounge. I raced to the kiosk and did a bulk purchase of the entire shelf of Sunlight brand bathing soap bars and Colgate brand toothpaste for US$40! This was to be my souvenir gift shopping for home!

Mexico—car breakdown and near holdup

On a business visit to Mexico in 1985, again to promote pre-fabricated shelter systems and housing, I took my family to Cancun to combine business with some pleasure.

On the second day, we planned to visit the famous archaeological site of Chichen Itza, which was some 200 kilometres away.

The day trip in a sightseeing bus would take us to the site and, with a touristy stopover at a market for shopping and lunch, return us by late afternoon. The adventurous streak had not left me. I rented a car, rationalizing that this would be an easy, two-hour, standard trip, and that we could be a little adventurous with the liberty of being independent and venture out into other attractions in the vicinity. This was to be a mistake.

We rented a small, old, four-door Toyota sedan, with Aruna sitting on the front passenger seat and my two young children on the backseat. We embarked on our trip to Chichen Itza after breakfast at the hotel. This was a straight drive, and driving conditions were excellent.

After going up the ancient steep steps of the pyramid, we went for lunch in the local market, did some more aimless sightseeing around the village, and started our drive back about late afternoon. An hour into the drive, the car started stalling, and I saw a barrier ahead that we had not seen on our drive to Chichen Itza. The barrier consisted of two tree trunks placed over empty oil drums with a short distance in between such that a motorist would be forced to slow down and make gentle turns around the barriers before continuing.

Miraculously, and resembling a scene as if it was choreographed for a movie, the car came to a stop right in the middle of this barrier, or so-called security check. Two young men sloppily dressed in combat gear and casually brandishing assault rifles, perhaps AK47s, emerged from the woods and headed for us.

Trouble lay ahead and, with the family in tow, seconds ticked like eternity. My mind was evaluating numerous factors. Why did we not see the barrier on our way out and who were these *"soldiers?"* Was this a hold up to rob us of our possessions? What if we resisted, or could we resist, and, if so, how would we escape in the wilderness as we were unfamiliar with the terrain? Would they shoot us after robbing us? None of us spoke Spanish. Where was the returning busload of tourists who had also gone to Chichen Itza? Had they left early or were they still to come? With these troubling thoughts, a decision had to be taken, and taken fast. By now, two more teenage *soldiers* wielding assault rifles walked over from the woods toward us. This was serious; it was not going to be easy, and big trouble loomed ahead.

An infantier's instinct flashed, and I decided to immediately take control and go on the offensive. I got out of the car very casually, opened the hood, oblivious to any threatening environment, and tried to fiddle around aimlessly to identify where the problem lay. I pretended I knew what I was doing. I showed no interest, fear, or concerns of the soldiers or if they indeed meant harm to us. By now, the four were standing and watching our every move, perhaps startled that I showed no fear of their presence, despite my heart throbbing at breakneck speed with worry.

With a strong loud voice, I shouted, expressing my remarks in a rambling mixture of the odd word of Spanish, Swahili, French, and perhaps Hindustani: *"Senor, senor, mucho problemo, tafadhali saidia, por favor, aide nous, rapido, rapido, jaldi, jaldi, etc."*

This went on for a few seconds and, whatever came to mind, I yelled out, taking special care not to display fear, but for whatever it was worth, expressing that I depended on them for help. One *soldier* came close, looked under the hood, and, bingo, as if he knew, fixed the problem. A wire from the distributor had come loose and he simply reconnected it. He asked me to try, and the engine fired. I thanked him loudly, *mucho gracias, senor, mucho gracias,* and gave him a traditional double handshake!

What a relief of immense proportions. But we were still at their mercy. By now, two more *soldiers* emerged, one of whom looked authoritative by the way he walked almost ceremonially. I greeted him with a traditional double handshake and gave him a bear hug. The other *soldiers* giggled and I knew that the situation was contained.

Now the bigger challenge was who to tip. The *soldier* who fixed the engine or their captain? Either way, it was not going to be easy. Whatever I gave would be considered too little and, just by opening my wallet in their presence, I was asking for trouble. If I gave to the *soldier* who had fixed the problem, the captain would feel bypassed and would expect a tip, also. And what about the other *soldiers* who were hanging around? They would also expect a tip, and I made the strong assumption that this roadblock was set up to rob tourists.

As a show of respect, I took off my baseball hat and placed it on the captain's head, got him giggling, and thanked him. Luckily, we all were wearing a baseball hat and had, by sheer luck, two more in our bag. I quickly placed one on each of the *soldiers*, repeating my thank-yous throughout, and left them giggling and laughing.

I got into the car, told everyone inside to open their windows and wave frantically as a mark of appreciation and profound thanks. I drove away very slowly, implying that we were not under any fear and, more importantly, knowing that the assault rifle had a range of over 300 metres and that we were not yet out of trouble. The car could still break down again.

Some milliseconds later, the road curved and, seeing through the rear-view mirror that we were out of their sight, my leaded foot pressed the accelerator all the way down just hoping that we would make it to Cancun before dark and without any further mechanical breakdown. Happily, we did, and whilst on the outskirts of the city, we noticed that a tourist bus was behind us in the distance. It was a relief of unimaginable proportions, and we were lucky to be out alive.

TRAVELS WITH JACQUES CHOUINARD

Jacques Chouinard, a popular Vandoos general from the famed 22nd Royal Regiment based in Valcartier, Quebec, had recently retired from his last job as commander, mobile command in Saint Hubert, Quebec. He had joined a company specializing in the manufacture of defence supplies.

He headed the marketing department with a mission to expand sales internationally. I had been representing the company for several countries in Africa, and he was my new point of contact.

We planned a marketing strategy whereby we would travel to Kenya, Tanzania, and Zambia to promote our products and other services. Travelling with him had bonuses. With a senior rank, a lot of doors at the very top opened effortlessly and we were able to get appointments with key officials with the help of the Canadian High Commissions easily. Our presentations to potential clients led to many lucrative contracts, and we repeated the visit a year later.

Whilst travelling and in each other's company for a long period, I bonded with him. He was always gracious and eloquent in his handling of people. We met cabinet ministers, heads of army and air force, and other senior bureaucrats. Business was conducted with ease. It was to our advantage that Canada was considered a neutral player in the ongoing Cold War that was slowly creeping into Africa; anything Canadian was considered good.

On one occasion in Dar es Salaam, Tanzania, we decided to invite the deputy head of police to dinner. The invitation was for 1800 hrs. I made reservations and selected a table for three overlooking the harbour with a lovely view. No one showed up. I did not have our guests' home phone numbers. We decided to wait for an hour, as I was familiar with a lacklustre respect for timing in the region. By 1900 hrs, still no one had showed up, and we had had our quota of pre-dinner vodka and tonics. We decided to go for dinner.

Just as we got seated, a young, beautiful girl, about twenty, approached us to say that her dad, the deputy, was on his way, and would be joining us shortly.

He had extended our invitation to his brother and a few close friends. The number arriving would be ten. With her, I quickly revised our reservation number to thirteen and got another table. Some twenty minutes later, our guest arrived with a contingent of twelve. The steward quickly added another table with some extra chairs in case we needed more. Jacques was getting uneasy all along. I had to calm his frustration, saying that it was not unusual for a guest to invite others without informing the host. A few more or fewer guests is not viewed as a big deal. We were in a precarious situation and had to remain calm, as promising business deals lay ahead, and, happily, we concluded them.

Jacques and his wife of over sixty years, Marriotte, became close family friends and we would interact socially on a regular basis. He would enlighten us on life in Canada with all its values, regional differences, political affiliations, and the never-ending issue of bilingualism and Quebec separatism.

The warmth with which he brought us into his circle of friends at the Citadel and the Garrison Club in Quebec City gave us enormous pride and comfort that it was now up to us to build our life in the new country. Such a warm welcome and acceptance as new immigrants, almost by circumstances by a senior, well-respected officer in our new country of adoption, gave us hope.

The opportunities were plenty, and people were there to help. It was up to us how we conducted ourselves. Jacques continued to have an impact on us until his untimely passing in 2010.

By now, I was well settled in my new occupation of an entrepreneur and expanded my interests to real estate, assembling a portfolio of impressive assets and aiming to go into development one day.

VOLUNTARY WORK, SPORTS IN SAINT BRUNO, QUEBEC

From my early days in Kenya, I had always been involved in voluntary work and have continued participating in several worthy causes. I assisted at the Dr. Bernardo's Homes and Starehe Boys' Centre, both organizations that gave shelter and hope for the homeless youngsters who had been discarded by society.

Firstly, when we moved to Saint Bruno on the outskirts of Montreal in 1980, I involved myself as a volunteer photographer with football—*Le Barons*—and basketball—*Blue Demons*—two local teenage teams, as our son, Sanjay, played both, and our daughter, Meera, played basketball and field hockey, the only female on a boys' team. Basketball was not yet an established sport in Saint Bruno. This was an enjoyable vocation working with enthusiastic kids but going against the traditional Quebec sport of ice hockey, about which I had very little knowledge.

I was reminded of this when I approached the mayor, Marcel Dulude, in 1986 for financial assistance to support our basketball team to participate in the Eastern Provincial Basketball Championships in Moncton, New Brunswick, where we would be representing Saint Bruno. His refusal to approve assistance was instant, maintaining that hockey is the sport in Quebec and not any other.

Disappointed with such a prompt refusal, I assembled a few parents and we drove the team in our cars to Moncton, some 800 kilometres away. Happily, after several matches, we were declared the runners-up and missed the championship by only a few points. The widespread press coverage meant Marcel Dulude received many congratulations. Our success put Saint Bruno on the map.

He called me on return several times to apologize for not having supported us, and to say he wanted to meet the winning team. He offered to present lapel pins to each of them in recognition for their efforts. We all knew that municipal elections were a few weeks away, that he wanted as much publicity as he could get, and that this was a perfect opportunity. We accepted his request, as it was more for the kids' pride. Both sports got a firm grounding in Saint Bruno, and continued to excel. Aruna and I were overjoyed that we had accomplished a feat against odds and that our involvement in the sports program had brought us closer to this small community. By now, we had overcome any worry about whether we had made the right decision to migrate to Canada. From the professional and social aspects, we were well-entrenched into a new society.

VOLUNTARY WORK— CANADIAN FORCES COLLEGE

My biggest voluntary support has been to the Canadian military, which I have considered as a continuation of my days in the KAR and the Kenya Army. I am a member and support the Royal Canadian Legion in Minden, Ontario. I became an associate member of the Officers' Mess, Canadian Forces College, Toronto, and offered to plan, organize, and execute several social events. One of these, the CornFest, an outdoor, country-style gathering of military personnel and their families, has become a major annual welcoming event at the start of the college year, running at present (2016) into its thirteenth year. I liaised with and got immense support from the Toronto Police Services, the Toronto Fire Department, and several volunteer veterans. We are also sponsors at the college, volunteering to act as points of contact for senior military officers from abroad who come to attend year-long courses. Such a close, ongoing involvement keeps us as a family, and especially me, closely connected to military colleagues, which I love enormously.

Remembrance Day Service at the Canadian Forces College, Toronto, reading Flander's Fields, 2009.

On another such voluntary support to the college, on a humanitarian situation of a gravely sick officer that I was asked to handle, I was recognized with the award of a Commandant's Commendation. I was also recognized by my federal Member of Parliament for my voluntary work with an award of the Queen's Diamond Jubilee Medal in 2012.

I have acted as a judge at the Warrior's Day Parade in Toronto at the opening of the annual event held at the Canadian National Exhibition. The parade consists of units of veterans from across North America and beyond who proudly march past in the original uniforms they wore in various wars and conflicts in the past, to much applause by spectators.

CANADIAN FORCES COLLEGE, TORONTO, THE SETHI INUKSHUK AWARD

In recognition of the immense opportunities Canada offered us, and with my love of the army, the family created an annual award in 2015, the *Sethi Inukshuk Award*. This is to be presented to a deserving student on the senior national securities program at the Canadian Forces College on graduation day in June for the next twenty years.

Each award is a white marble *inukshuk,* hand-made by a distinguished, well-known artist, and mounted on a black marble base, encased in a natural pine presentation box. This initiative was supported by the college and will remain as one of the key awards for many years.

The *inukshuk* is a very Canadian stone monument erected in the shape of a human and is widely used as a direction marker in the harsh and desolate Arctic wilderness.

It is a tool of survival and symbolizes co-operation and unselfishness for the greater good of the group over the individual. It was selected for its suitability and appropriateness as it symbolizes a survival instinct to remain focussed, much as has been the case for me in pursuit of my goals, to *shape my destiny.*

The Sethi Inukshuk Award, created 2015

Meeting Governor General David Johnston, at the
Canadian Forces College, Toronto, 2012

Aruna and I with Prime Minister Stephen Harper, Toronto, 2014

Medals; left to right: Distinguished Service Medal, Shifta Campaign Medal, General Service Medal and Queen's Diamond Jubilee Medal, 2012

CANADA, A HOME FOREVER

Since moving to Canada, we have immersed ourselves fully into a new society, by making adjustments to our lifestyle and celebrating all the cultural and national events. We have blended in very well, and are proud to call ourselves Canadian in this multicultural mosaic where people in our neighbourhood, at work, and elsewhere have ancestral roots outside this country. We have also enriched a new society with many contributions. We have never questioned whether our choice to move to Canada was the right one. Canada is now home, and will remain as such for my family, children, and grandchildren.

Having started my business from scratch with huge potential to expand and grow, Aruna and I hope that our children will take it to the next level. We continue to instill in our family, extended family, and friends that there are obligations and responsibilities that go with citizenship in a new country. We voluntarily chose to make our home in Canada, and have benefitted from assistance and support from federal, provincial, and municipal levels of government, along with several people around us. It is now our unreserved duty to pay back voluntarily at our own initiative to others who are in need and have similar ambitions to succeed as we did.

And just as we have been fortunate to receive inspiration and guidance from others, we feel it is our commitment to provide such assistance wherever we can to others. Sharing with others what we have is fundamental to our outlook.

Canada has enriched us with pride, opportunities, and hope for the future. Our interest now is to encourage our grandchildren to serve Canada in uniform, either as reservists or regulars. We hope they will sustain a military tradition in the family and, most importantly, will play their part to protect our freedom for which many brave went to distant lands to fight wars to keep us safe, but sadly did not return to enjoy what they gave us so freely.

CONCLUSION

My colourful journey over the years has been most satisfying and fulfilling with an assurance that, within reason, any goal is attainable. Several factors played a key role in my success, such as identifying a direction, setting a goal, and preparing to fulfil an ambition, thereby *shaping my destiny*.

A constant reminder has been to share a surplus with those in need, as we all have surpluses of all sorts, the most important being the time to help others. No one can go through life singlehandedly and alone. The oft-heard phrase that "it takes a village to raise a child" is so very true. I have been a beneficiary of such gratitude from friends, colleagues, and strangers whose paths I've crossed along the way.

I have often wondered how I managed to succeed from humble beginnings by starting from scratch and running a very successful business venture in a new country. Some people with the benefit of an inheritance or support from a well-established family with the right connections would simply walk through an open door to a waiting chair with no effort whatsoever. I think such an approach where one has not earned his stripes but has been privileged to get everything on a silver platter with no effort often loses the value of a possession. It is important for one to know what it takes to get ahead in life or what the real value of one's possessions is, rather than just the price. Most importantly, they forget that relations are a two-way street; you get what you put in. Complacency sets in when you get something for nothing, and it is a given that, when you are making the most with the least, it promotes initiative that leads to creativity, thus giving you pride of immeasurable proportions for being self-made.

It is all very well to be shown the right door where an opportunity lies, but often an opportunity is discovered when that door remains closed and other options have to be found or created to get across. It takes an awful lot of courage and willpower to satisfy yourself that you have the capacity to take on a challenge head on. It is no different than writing a resume for a job application, where every effort is made to sparkle the positive characteristics for a stranger to take notice and evaluate them for compatibility.

The most important factor here is for you as the person to feel convinced that you have what it takes to fulfil a dream. It has to come from within.

In my case, I had to develop a vision, set achievable goals, and remain focussed. I continued seeking guidance, counsel, and advice from those I trusted and counted among my small but growing influential circle of confidantes.

To achieve these goals, I had to have an education and to qualify as a candidate in whatever career goal I pursued. Like a road map, I had to chart out a direction to my destination with the full knowledge that the road would be bumpy with forks, obstacles, and disappointments along the way. I had to remain focussed with a clear aim to pursue my vision with the best of my ability. I discovered early on that my families' financial stability in my youth would have been desirable, but I never felt that I was a step below others or considered a financial need beyond basics as a shortcoming, if it can be termed a shortcoming. In the bigger picture, I more than compensated for any financial misgivings with hard work, dedication, integrity, and a relentless drive to succeed. This is all that was required to gain the trust and respect of others, and to slowly build a reputation of integrity, loyalty, and reliability.

Tracing my early years of learning, I can say that whatever was taught under a tree out in the open by any elder willing to impart any knowledge was as much a part of my education as any I received in a formal classroom setting. Such an environment fostered self-reliance, and matured me at a young age where I would be several steps ahead of my peers. This was not a case of missing out on childhood; on the contrary, it was to ignore childish and inconsequential things or toys which, in some way, were meant to give temporary pleasure and a false sense of ownership.

It was not an easy decision to forego a full scholarship to pursue a career in engineering with a promising and lucrative future ahead, but that was not my calling. My relentless drive to pursue a career in the army, at times fearing that I was out of my depth but always raising the bar to constantly strive for the best, formed a part of my personality. I will always rate attending Sandhurst and Camberley on merit as a great achievement.

The challenges that I encountered at a young age commanding troops with a wide range of beliefs, traditions, cultures, language, religion, and aptitude provided a mixture of education and understanding of people.

During my first deployment in field, the success I achieved in moulding a diverse body of men in a short space of time into a fully fledged fighting force by leading troops in harm's way transcended every imaginable barrier. Such a goal did not come easily, neither it was followed from a textbook. With clarity of vision, setting an achievable goal, determination, and full respect for my men, I earned their respect and loyalty willingly.

I did not have to resort to threats, special treatment for some, or pretending to overlook where taking a hard decision could have invited unpleasantness. My treatment of my men was always firm and fair, following an established tradition of military discipline. Yet when flexibility allowed me to handle minor infractions, I resorted to a chat rather than throwing the book at the accused. It worked, and continued to earn me loyalty.

The human factor is all-important to whatever we do since the biggest joy is to blend one's resources into a collective effort, to complement achievement, and to identify and admit a failing.

Reputation, like a shadow, follows one at whatever speed one goes, and a track record is the yardstick of measuring performance. My interaction at many a crossroad with a vast range of experienced people was sustained with care, trust, and sincerity. There was never any discomfort for me that going through strange surroundings, I was the odd one, whether from a racial perspective or perhaps because I was the youngest in command. Throughout my army career, in both the field and at headquarters, I handled assignments and tasks at a national level, and such responsibilities were entrusted to my care by my superiors with the confidence that I would perform to, or exceed, expectations. I handled these responsibilities at a captain and major rank, which are presently held by a brigadier or major general rank.

If I did not know something, I asked. Rank and authority are fundamental in a chain of command in any organization and, beyond that, it is the people who form and sustain such links. The biggest asset in any organization is human, and no matter how good a plan, it is the people around you who will implement it, and they will either exceed expectations or fall short, depending on where their heart is. As humans, we have strengths and weaknesses of all kinds, and the key is to draw from the strengths of others and complement their weaknesses wherever it is possible.

I owe my success to the extraordinary, larger-than-life people I was fortunate to have met and in whom I confided. The Duke of Norfolk and Sir Anthony Duff were people destined for a place in history, and they moved in exclusive, top echelons of society where access to them for the majority was not possible. But here I was, at the bottom of the totem pole, enjoying access, respect, and their counsel because they wanted to help me. Their generosity in assisting and guiding me has been phenomenal and, in retrospect, I guess they must have identified some qualities in me beyond sincerity, loyalty, and team playership.

Such support came in at no financial cost, and I could never put a dollar amount to what I achieved—and which I could never ever pay back in monetary terms. Hence, my goal, repeated time and again to whomever I come across, is to share their surplus and bring hope to those seeking to better themselves.

You have to have a vision to succeed: an *ambition,* which has to come from within; *determination* to remain focussed; and *leadership* to chart out a clear direction from a mass of conflicting options. The collective effort from those around you to willingly support you is fundamental to *shaping a destiny.*

REFERENCES AND BIBLIOGRAPHY

The King's African Rifles by Lieutenant Colonel Moyse-Bartlett

A study in the military history of East and Central Africa, 1890-1945

A History of The King's African Rifles by Malcolm Page

Wikipedia